T0230824

Proceedings
from the Medical Workshop
on Pesticide-Related Illnesses
from the International Conference
on Pesticide Exposure and Health

Proceedings from the Medical Workshop on Pesticide-Related Illnesses from the International Conference on Pesticide Exposure and Health has been co-published simultaneously as *Journal of Agromedicine*, Volume 12, Number 1 2007.

Proceedings from the Medical Workshop on Pesticide-Related Illnesses from the International Conference on Pesticide Exposure and Health

Ana Maria Osorio, MD, MPH
Lynn R. Goldman, MD, MPH
Editors

Proceedings from the Medical Workshop on Pesticide-Related Illnesses from the International Conference on Pesticide Exposure and Health has been co-published simultaneously as *Journal of Agromedicine*, Volume 12, Number 1 2007.

Routledge
Taylor & Francis Group

LONDON AND NEW YORK

First published 2007 by The Haworth Medical Press®

2 Park Square, Milton Park, Abingdon, Oxon OX14 4RN
711 Third Avenue, New York, NY 10017, USA

Routledge is an imprint of the Taylor & Francis Group, an informa business

First issued in hardback 2017

Proceedings from the Medical Workshop on Pesticide-Related Illnesses from the International Conference on Pesticide Exposure and Health has been co-published simultaneously as *Journal of Agromedicine*, Volume 12, Number 1 2007.

Copyright © 2007 Taylor & Francis

All rights reserved. No part of this book may be reprinted or reproduced or utilised in any form or by any electronic, mechanical, or other means, now known or hereafter invented, including photocopying and recording, or in any information storage or retrieval system, without permission in writing from the publishers.

Notice:
Product or corporate names may be trademarks or registered trademarks, and are used only for identification and explanation without intent to infringe.

Library of Congress Cataloging-in-Publication Data

Medical Workshop on Pesticide-related Illnesses (2002 : Washington, D.C.)
     Proceedings from the Medical Workshop on Pesticide-related Illnesses from the International Conference on Pesticide Exposure and Health / Ana Maria Osorio, Lynn R. Goldman, editors.
          p. ; cm. – (Journal of agromedicine ; v. 12, no. 1)
     "Based upon presentations by the authors at the 2002 International Conference on Pesticide Exposure and Health in Washington, D.C."–Pref.
     Includes bibliographical references and index.
     "Published simultaneously as Journal of agromedicine, Volume 12, Number 1, 2007.
     ISBN 978-0-7890-3578-3 (soft cover : alk. paper)
     1. Pesticides–Toxicology–Congresses. 2. Pesticides–Environmental aspects–Congresses. 3. Environmentally-induced diseases–Congresses. I. Osorio, Ana Maria. II. Goldman, Lynn R. III. International Conference on Pesticide Exposure and Health (2002 : Washington, D.C.) IV. Journal of agromedicine. V. Title. VI. Series.
     [DNLM: 1. Pesticides–toxicity–Congresses. 2. Environmental Exposure–adverse effects–Congresses. 3. Environmental Health–methods–Congresses. W1 JO534JD v.12 no.1 2007 / WA 240 M489p 2007]
RA1270.P4M42 2002
363.738'498–dc22

                                                                                                                2007012990

ISBN 978-0-7890-3578-3 (pbk)
ISBN 978-1-138-45529-0 (hbk)

Proceedings
from the Medical Workshop
on Pesticide-Related Illnesses
from the International Conference
on Pesticide Exposure and Health

CONTENTS

ABOUT THE EDITORS

Ana Maria Osorio, MD, MPH, is a Medical Officer in the U.S. Public Health Service and currently the Regional Medical Officer for the Pacific Region of the U.S. Food and Drug Administration. She is board-certified in Occupational Medicine and has an MPH in Epidemiology. Her training includes serving as an Epidemic Intelligence Service Officer at the National Institute for Occupational Safety and Health (NIOSH), Centers for Disease Control and Prevention (CDC). Prior to her current position, Dr. Osorio served within the California Department of Health Services (as Chief of the Occupational Health Branch, and later as Chief of the Division of Environmental and Occupational Disease Control), and, subsequently, acted as Medical Officer for the Office of Pesticide Programs at the U.S. Environmental Protection Agency. Dr. Osorio is the author of numerous book chapters and journal articles dealing with various aspect of environmental and occupational medicine. Her research topics include agricultural worker health, pesticide-related health conditions, reproductive hazards in the environment, ergonomic risk factors, environmental pulmonary hazards, tribal medicine, cancer workplace risks, adult and child lead intoxication, public health disease monitoring systems, and international public health issues.

Dr. Osorio has served on numerous federal advisory committees and task force groups which deal with environmental and occupational health issues; the topics covered by these groups include asthma, health care providers hazards, field applications for environmental biomonitoring, health and safety in the construction industry, scientific overview of activities at the Agency for Toxic Substances and Disease Registry (ATSDR), energy-related epidemiologic research at the Department of Energy (DOE) and CDC, national pesticide illness and injury reporting, childhood agricultural injuries, environmental medicine training among health care providers, migrant and seasonal worker health, environmental and occupational illness at DOE sites. In addition, Dr. Osorio has lectured worldwide on international environmental and occupational health issues.

Dr. Osorio's current work at FDA involves providing medical assistance for Pacific Region disease outbreaks for FDA regulated products, developing training on food security and counter-terrorism, creating training curricula on the evaluation of disease outbreaks and related epidemiological concepts, and providing consultation to the regional staff on medical and epidemiological issues. Through her Public Health Service assignments, Dr. Osorio has been involved in emergency response efforts for hurricane and typhoon-related events.

Lynn R. Goldman, MD, MPH, is professor of environmental health sciences, health policy and management for the Johns Hopkins Bloomberg School of Public Health in Baltimore, Maryland.

One of the most influential public health physicians of our time, Dr. Goldman speaks for those members of our society who rarely have a voice in policy-making decisions–our children. A pediatrician and epidemiologist, Dr. Goldman combines her two specialties to improve national health policy, especially children's environmental health. Between 1985 and 1992, Dr. Goldman served in various positions in the California Department of Health Services, most recently as Chief of the Division of Environmental and Occupational Disease Control, where she was responsible for the conduct of a number of epidemiological investigations of the impacts of environmental exposures to health, especially the health of children. In 1993, Dr. Goldman was appointed by President Clinton and confirmed by the Senate to serve as Assistant Administrator for Prevention, Pesticides and Toxic Substances at the U.S. Environmental Protection Agency (EPA). In that position, she was responsible for the nation's pesticide and toxic chemicals regulatory programs at the EPA. Her efforts resulted in the Food Protection Act passed by Congress in 1996, the first national environmental law to explicitly require measures to protect children from lead poisoning and pesticides. In January 1999 she left the EPA and joined Johns Hopkins University.

Dr. Goldman has served on numerous national boards and expert committees including the Committee on Environmental Health of the American Academy of Pediatrics, the Centers for Disease Control Lead Poisoning Prevention Advisory Committee and the National Research Council.

Preface:
Pesticide-Related Illnesses

The *Journal of Agromedicine* is very pleased to have Ana Maria Osorio, MD, MPH, and Lynn R. Goldman, MD, MPH, serve as guest editors of a special two-part issue on pesticide-related illnesses. Part II, scheduled for publication at a later date, will address pulmonary, children's health, cardiac, medical surveillance and environmental history taking issues.

The papers are updated and expanded reviews based upon presentations by the authors at the 2002 International Conference on Pesticide Exposure and Health in Washington, DC. All of the authors have revised their presentations to address the current status of knowledge in pesticide health effects upon the major body organ systems, including the role of pesticides in carcinogenesis. Our goal is for the special issue to serve as a reasonably-priced current reference on pesticide related illnesses.

In addition to the review papers, the issue includes appendices containing annotated Web-based references, acute poisoning severity rating scales, and reporting documents. These additional appendices can be used to document and report pesticide poisonings in a manner consistent with World Health Organization (WHO) international standards. Drs. Goldman and Osorio together have impeccable credentials to serve as guest editors for this issue. They have been involved with pesticide-related health effects from academic, state and federal regulatory, and public health research perspectives. Contributing authors were chosen for their body of experience in both research and/or consulting clinical practice involving pesticide exposures.

Previous papers in the *Journal of Agromedicine* and other journals have addressed the lack of knowledge and training of both urban and rural clinical health care providers and public health practitioners in the discipline of pesticide toxicology.[1] We hope to make the peer-reviewed current references available to aid in the ongoing effort to improve health care provider knowledge in this topic area. An ongoing lack of understanding of the role of pesticides in modern day agriculture, and the public health impact by both the lay public and health practitioners, continues to exist, and scientifically valid information is scattered or inaccessible due to the cost of reference texts. Notification of the publication date of the complete monograph incorporating both issues will be forthcoming.

Steven R. Kirkhorn, MD, MPH, FACOEM

REFERENCE

1. Kirkhorn SR. Editorial. J Agromedicine. 2006; 11(3/4):1-3.

[Haworth co-indexing entry note]: "Preface: Pesticide-Related Illnesses." Kirkhorn, Steven R. Co-published simultaneously in *Journal of Agromedicine* (The Haworth Medical Press, an imprint of The Haworth Press, Inc.) Vol. 12, No. 1, 2007, pp. xxi; and: *Proceedings from the Medical Workshop on Pesticide-Related Illnesses from the International Conference on Pesticide Exposure and Health* (ed: Ana Maria Osorio, and Lynn R. Goldman) The Haworth Medical Press, an imprint of The Haworth Press, Inc., 2007, pp. xv. Single or multiple copies of this article are available for a fee from The Haworth Document Delivery Service [1-800-HAWORTH, 9:00 a.m. - 5:00 p.m. (EST). E-mail address: docdelivery@haworthpress.com].

Available online at http://ja.haworthpress.com
© 2007 by The Haworth Press, Inc. All rights reserved.

Introduction/Background

Lynn R. Goldman, MD, MPH
Ana Maria Osorio, MD, MPH

Pesticides are chemical, biologic and other types of agents used to kill or inhibit the growth of pests of economic importance. The target function of the pesticide is reflected in the following agent categories: insecticides, fungicides, herbicides, rodenticides, wood preservatives and disinfectants. The type of pesticide may include examples of synthetic or inorganic chemicals, metals, genetically modified plants and biological materials (such as pheromones, and *Bacillus thuringiensis* bacteria). As such, pesticides provide benefits to human welfare by contributing to agricultural productivity, public health pest control and industrial production. At the same time, pesticides have been of particular concern for health because of their biological activity, especially acute human health effects. In consequence, the U.S. Environmental Protection Agency (EPA) has published a handbook of pesticide poisonings that provides information about the acute toxicity of pesticides and how to identify and treat intoxications.[1]

Prevention of acute pesticide intoxication is an important issue worldwide. However, the lesser known pesticide-related chronic health hazards are of at least equal importance. There are many challenges to understanding the chronic health effects of pesticides, such as the hundreds of pesticide active ingredients and other ingredients that part of the product formulations, the nature of the temporal association between pesticide exposure and chronic health effects (due to latency between exposure and clinically detectable effects), and the absence of biological markers for chronic exposures for most pesticides. Nonetheless, there is much that is known about the chronic pesticide-related effects. For this reason, this special issue of the *Journal of Agromedicine* provides a timely review of both the acute and chronic effects of pesticides on human health.

The workshop titled "Pesticides and National Strategies for Health Care Providers," held in Washington, DC, in April 1998, formed the genesis for this compilation of articles on pesticide-related medical effects. The goal of the workshop was to identify strategies for educating health care providers on the recognition, management and prevention of pesticide intoxications. Key governmental and non-governmental stakeholders attended the event, which

Lynn R. Goldman is Professor, Environmental Health Sciences, Johns Hopkins Bloomberg School of Public Health, Baltimore, MD. Ana Maria Osorio is Regional Medical Officer for the Pacific Region of the U.S. Food and Drug Administration.

Address correspondence to: Ana Maria Osorio, MD, MPH, U.S. Food and Drug Administration–Pacific Regional Office, 1301 Clay Street, Suite 1180N, Oakland, CA 94612 (E-mail: anamaria.osorio@fda.hhs.gov).

Disclaimer: Drs. Goldman and Osorio initiated their work on this project while at U.S. Environmental Protection Agency, Washington, DC. The opinions and contents of this article are those of the authors and do not necessarily reflect the views of the U.S. Food and Drug Administration, nor the U.S. Environmental Protection Agency.

[Haworth co-indexing entry note]: "Introduction/Background." Goldman, Lynn R., and Ana Maria Osorio. Co-published simultaneously in *Journal of Agromedicine* (The Haworth Medical Press, an imprint of The Haworth Press, Inc.) Vol. 12, No. 1, 2007, pp. 1-2; and: *Proceedings from the Medical Workshop on Pesticide-Related Illnesses from the International Conference on Pesticide Exposure and Health* (ed: Ana Maria Osorio, and Lynn R. Goldman) The Haworth Medical Press, an imprint of The Haworth Press, Inc., 2007, pp. 1-2. Single or multiple copies of this article are available for a fee from The Haworth Document Delivery Service [1-800-HAWORTH, 9:00 a.m. - 5:00 p.m. (EST). E-mail address: docdelivery@haworthpress.com].

Available online at http://ja.haworthpress.com
doi:10.1300/J096v12n01_01

launched the National Strategies for Health Care Providers Pesticide Initiative. This initiative was created by the EPA and the National Environmental Education and Training Foundation in collaboration with the U.S. Departments of Health and Human Services, Agriculture and Labor. The goal is to improve the diagnosis, management and prevention of pesticide-related intoxications by health care providers. To date, seven federal agencies and 16 health care professional associations have been involved in the development of this initiative.[1]

EPA has continued this activity via cooperative agreements with the Migrant Clinicians Network (MCN) and the Pacific Northwest Agricultural Safety and Health Center (PNASH). The goal of the MCN project is to work directly with the health care community to improve pesticide education and awareness with special emphasis on the training of health care providers in the recognition and treatment of pesticide-related health conditions. The MCN investigators will test a training model for primary health care providers that will incorporate key practice skills. The PNASH project will incorporate the recognition and management of pesticide intoxications into the education of future health care providers. This will be accomplished by integrating the Initiative's core competencies into the curricula of medical, nursing, and physician assistant training programs in the Pacific Northwest. Both of these projects will be evaluated for possible application in other regions of the country.

In July of 2002, the International Conference on Pesticide Exposure and Health was one event that came out of this Initiative. This event was held in Bethesda, Maryland, hosted by the Society for Occupational and Environmental Health (SOEH) and co-sponsored by EPA, the U.S. Centers for Disease Control and Prevention (CDC) and the National Institute for Environmental Health Science (NIEHS). Included in this conference was a two-day medical workshop for health care providers which showcased presentations on pesticide-related illnesses. The topics covered during this medical workshop have been updated and are now included as articles in this special issue. Our intent is to provide a timely review of key issues in both the acute and chronic health effects of pesticide exposure that health care and public health practitioners will find useful.

REFERENCE

1. Available from: www.epa.gov/oppfead1/safety/healthcare/healthcare.htm [cited 2006 Nov 23].

Contact Dermatitis in Agriculture

Sahar Sohrabian, MD
Howard Maibach, MD

SUMMARY. Dermatitis in agricultural workers, albeit a common entity, infrequently receives scrutiny as to mechanism and prevention. We briefly describe potential mechanisms, examples for each, and offer an algorithm for etiologic diagnoses. doi:10.1300/J096v12n01_02 *[Article copies available for a fee from The Haworth Document Delivery Service: 1-800-HAWORTH. E-mail address: <docdelivery@ haworthpress.com> Website: <http://www.HaworthPress.com> © 2007 by The Haworth Press, Inc. All rights reserved.]*

KEYWORDS. Irritant dermatitis, allergic contact dermatitis, agricultural chemical

INTRODUCTION

Contact dermatitis is a skin reaction resulting in erythema and edema, and sometimes vesicles and bullae, due to an inflammatory response to an antigen or an irritant. Late manifestations may include weeping, crusting, secondary infection, and post-inflammatory hypo and hyperpigmentation. The main symptom is pruritus. There are different causes and mechanisms of contact dermatitis; however, the resulting inflammatory response often looks similar. Contact dermatitis can be divided into acute, subacute, and chronic contact dermatitis. In an acute contact dermatitis, the skin may be red and edematous and vesicular, whereas in a subacute reaction, papules and edema may be the dominant lesion. In chronic contact dermatitis, the skin has may have less edema and instead present with more scaling and lichenification. Mechanisms include allergic, photoallergic, irritant, photoirritant, and contact urticaria. This overview focuses on dermatitis and urticaria secondary to agricultural chemicals.

Irritant Contact Dermatitis

Irritant dermatitis is an inflammatory reaction resulting from an irritating agent. Irritant contact dermatitis is often caused by acids, alkalis, or solvents. The severity of the skin lesions produced by the irritants is related to the amount of time exposed, the dose, and chemicals potency. Patients who already have endogenous eczema are more likely to develop irritant contact dermatitis when the irritant comes into contact with their skin. Types of irritant

Sahar Sohrabian and Howard Maibach are affiliated with the University of California-San Francisco, Dermatology Department.

Address correspondence to: Howard Maibach, MD, University of California-San Francisco, Dermatology Department, P.O. Box 0989, 90 Medical Center Way 110, San Francisco, CA 94143-0989 (E-mail: maibachh@ derm.ucsf.edu).

[Haworth co-indexing entry note]: "Contact Dermatitis in Agriculture." Sohrabian, Sahar, and Howard Maibach. Co-published simultaneously in *Journal of Agromedicine* (The Haworth Medical Press, an imprint of The Haworth Press, Inc.) Vol. 12, No. 1, 2007, pp. 3-15; and: *Proceedings from the Medical Workshop on Pesticide-Related Illnesses from the International Conference on Pesticide Exposure and Health* (ed: Ana Maria Osorio, and Lynn R. Goldman) The Haworth Medical Press, an imprint of The Haworth Press, Inc., 2007, pp. 3-15. Single or multiple copies of this article are available for a fee from The Haworth Document Delivery Service [1-800-HAWORTH, 9:00 a.m. - 5:00 p.m. (EST). E-mail address: docdelivery@haworthpress.com].

Available online at http://ja.haworthpress.com
© 2007 by The Haworth Press, Inc. All rights reserved.
doi:10.1300/J096v12n01_02

contact dermatitis include acute and cumulative irritant contact dermatitis. In acute irritant contact dermatitis, after a chemical substance contacts the skin there is almost immediate erythema, blistering, and ulceration. In contrast, in cumulative irritant dermatitis, the skin needs to be exposed to the chemical on multiple occasions before erythema, dryness and thickening, and painful fissuring occurs. Chew provides extensive detail on current knowledge of the many types of irritation.[1]

Allergic Contact Dermatitis

In allergic contact dermatitis, a cell-mediated type IV delayed hypersensitivity reaction predominates. A single exposure or repeated contact with an allergen may induce sensitization. Once T lymphocytes are sensitized and an allergy has developed, skin lesions develop with within approximately 1-7 days post exposure. At the site of skin contact, an erythematous and often edematous eruption usually develops which then may form vesicles and bullae. A few days following the primary lesions, dermatitis at sites that are unrelated to the contact site may appear due to transfer of the chemical from the hands. The most common causes of allergic contact dermatitis in the USA include poison ivy, oak, or sumac. However other substances that can produce a similar reaction include nickel sulfate, potassium dichromate, mercaptobenzothiazole, and thirams. Patch testing helps identify a lesion as a result of allergic contact or photoallergic dermatitis.[2]

Photodermatitis (Photoirritation [Phototoxicity] + Photoallergic Contact Dermatitis)

In photodermatitis, an irritant or allergen is deposited in the skin and then typically activated by UV light in the range of 310-430 nm. Thus, the irritant can become phototoxic. A skin lesion produced by phototoxicity appears as severe sunburn. The skin is erythematous and may heal with hypo- and/or hyperpigmentation. Agricultural phototoxic dermatitis reactions often result from exposure to plants such as celery, lime and parsnips. In addition, wild parsnip and gasweed contain the furocoumarin that causes the phototoxic reaction. Medications

such as sulfa drugs, thiazides, and tetracycline–particularly doxycycline–are phototoxins that form a more homogenous skin lesion pattern on the area of skin affected. An immunologically mediated reaction (photoallergic contact dermatitis) is less common than the skin reactions caused by phototoxicity.[3]

Contact Urticaria (Non-Immunologic [NICU] + Immunologic [ICU])

A type I IgE-mediated process mediates immunologic contact urticaria and results in an immediate wheal and flare reaction. A strong contact urticant allergen or high dose of high clinical sensitivity may cause systemic symptoms such as asthma and anaphylactic shock. One agent identified to cause immunologic contact urticaria is the insect repellent diethyltoluamide (DEET).

Non-Immunologic contact urticaria (NICU) is more common than the immunologic form (ICU) and only produces local edema and erythema at the affected sites and not systemic effects; distal organs are not involved. The offending agents include benzoic, sorbic, or nicotinic acids.[4]

Diagnostic Testing

Although irritant contact dermatitis is the most common form of contact dermatitis, and skin lesions can form immediately after first time contact with an irritant, there is no diagnostic test. Irritant contact dermatitis is currently a diagnosis of exclusion, often aided by negative patch testing to rule out allergic contact dermatitis. Evaluate the patient by obtaining a detailed history of the chemicals and the amount of exposure the patient has encountered. In agriculture the agents that are encountered include pesticides, fertilizers, and secretions from animals.

Allergic contact dermatitis is caused by a delayed hypersensitivity reaction. The skin must be previously exposed to the allergen and then with future contact the skin forms an erythematous, edematous, and vesicular reaction. Patch testing, intradermal tests, or open tests can help diagnose allergic contact dermatitis; patch testing is the most studied of these methods.

Two patch test methods to diagnose allergic contact dermatitis include the TRUE test and Finn Chamber method. The TRUE test is previously prepared and the strips are simply applied to test for the suspected allergen. Unfortunately, TRUE test for agricultural chemicals is not available. Substances not tested by the TRUE test are covered by the more flexible test, the Finn-Chamber method. In the patch test, the allergen is placed on a chamber on adhesive tape and then adhered to the skin. The patch remains for 48 hours, and readings are at 49 and 96 hours. A positive patch test shows a skin area of erythema, edema, and often vesicles. In an allergic reaction, the skin response may extend outside the patch boundaries, whereas an irritant reaction typically has skin manifestations that peak within 48 hours and then fades. The limitations of patch testing include false negative results if there is insufficient penetration of the allergen. False negatives may also be produced if the patch testing is done on the lower back or forearms. A false-positive result may appear if the test compound is at an irritant level, or the individual has skin inflammation elsewhere. Another limitation of the testing is that patients with allergic eczema may experience transient exacerbations.

Patch testing has largely replaced intradermal testing. Intradermal testing is used in situations when allergic contact dermatitis is highly suspected but patch testing is negative. Intradermal testing may be more useful than patch testing in hydrocortisone contact allergy. The drawbacks of intradermal testing include that it is an invasive procedure, and it may also cause anaphylactic and general skin reactions. The materials used in an intradermal test include allergens in a vehicle, which is usually normal saline delivered through tuberculin syringes. To administer the intradermal test, .05-.1 ml of allergen is intradermally applied to the volar aspects of the forearms. The intradermal test is read at 30 minutes, 24 and 48 hours. An erythematous and edematous skin reaction at 30 minutes implicates an immediate type I allergy. An arthus reaction with polymorphonuclear leukocyte infiltration after 2-4 hours is more suggestive of a cytotoxic type III reaction. A delayed reaction (24 hours +) points to delayed hypersensitivity.

However, when a patch test is negative, but the history strongly suggests contact allergy, an open use test may be conducted. In an open test, the putative allergen is applied in a vehicle such as ethanol, or water, the actual product, and an applicator is used to spread the allergen typically on the antecubital fossa for up to 28 days or until a positive reaction occurs. A positive reaction consists of erythema and edema. Villarama provides details.[5]

Photoallergic contact dermatitis is suspected by the clinical presentation at photoexposed sites of the skin, but the allergy is confirmed by photopatch testing. In the photopatch test, two sets of patch tests are applied to the skin for 24-48 hours. At 24 hours one patch set is irradiated with UVA at a dose below minimal erythematous dose (5-10 joules). There is no UV dose applied to the other set. The reading time for the photopatch test is at 48 and 96 hours. If, after the readings, there is only a reaction at the irradiated site, then the reading indicates a photoallergy. However, if a reaction happens at both sites, including the non-irradiated site, this suggests contact allergy and not photoallergy.

When symptoms such as pruritus, tingling, erythema, and a wheal and flare reaction appear minutes after contact with a substance, this is referred to as a contact urticaria syndrome. Contact urticaria can be either immunologic (type I, IgE mediated, ICU) or non-immunologic contact urticaria (NICU). Two types of diagnostic tests are an open test and a prick test. In an open test, an allergen in a vehicle often (ethanol or water) is applied to the skin. The skin reaction read up to 1 hour consists of erythema and/or urticara. The open test is diagnostic for NICU and ICU, in that a reaction indicates contact urticaria and does not require controls.

In a prick test, one drop of each test allergen and histamine control is applied to the volar aspect of the forearms; a lancet is used to pierce the test site so the allergen can be introduced into the skin. The test is read after 15-30 minutes. A positive test is when an edematous reaction of at least 3 mm in diameter and half the size of the histamine control appear. Controls are necessary to rule out non-specific reactions. NICU such as with DMSO will occur in almost all individuals; ICU occurs only in the sensitized individual.

Methods for Testing the Irritation and Sensitization Potential

Assaying skin irritation is done in albino rabbits (Draize model). The Draize test measures irritation by averaging the erythema and edema scores at all sites. The test is quantified into a PII value. If the score is < 2 it is non-irritating, 2-5 mildly irritating, and > 5 severely irritating. Rabbits are more sensitive to irritants than the human skin.

In a cumulative irritation assay, the effects of cumulative exposure to a potential irritant is studied. Frosch et al. described the guinea pig repeat irritation test (RIT) to evaluate protective creams against the chemical irritants sodium lauryl sulfate, sodium hydroxide, and toluene.[6] The irritants were applied daily for two weeks to shaved back skin of young guinea pigs. Barrier creams were applied to the test animals 2 hours prior to, and immediately after, exposure to the irritant. Irritation was measured by visualizing the erythema. Marzulli and Maibach described a cumulative irritation assay in rabbits, which utilizes open applications and control reference compounds.[7] The test substances are applied 16 times over a three-week period. Erythema and skin thickness are used to measure resultant irritation.

The mouse ear model is used for simplicity, however, relevance to human findings remains unclear. Uttley and Van Abbe applied undiluted shampoos to one ear of mice daily for four days, visually quantifying the degree of inflammation as vessel dilatation, erythema, and edema.[8] Patrick and Maibach measured ear thickness to quantify the inflammatory response to surfactant-based products and other chemicals.[9]

In human model tests (Draize Repeat Insult Patch Test), the test area is small and localized, therefore several products can be tested simultaneously and compared. A reference irritant may be included to compare the test substances and examine their irritant capacity.

The National Academy of Sciences reviewed skin irritation in humans through a single application patch test procedure. The occlusive patches may be tested on the intrascapular region of the back or the dorsal surface of the forearms. A four-hour exposure period is suggested for the occlusive patch test, however for a volatile material a 30-60 minute period may be used for test. The test is evaluated by the appearance of erythema on the skin.

Lanman et al. and Phillips et al. described the 21-day cumulative irritation assay. In this test, a 1-in piece of Webril soaked with liquid is applied to a pad.[10,11] This site is examined after 24-hour after which the pad is reapplied. This process of examination and reapplying the pad to the skin continues for 21 days. The chamber scarification test is similar, however damaged skin is being studies versus healthy skin.

In human allergy tests, the material being tested is applied to the skin under a patch. The induction sensitization is repeated approximately nine times over a three-week induction period. Then there is a 10-14 day rest period. After the rest period a non-irritating dose is applied to the skin for the challenge test.

Finally, in quantitative structure analysis relationships, physiochemical data gathered on identified sensitizers is used to determine if substances being tested are sensitizers as well.

Skin Manifestations from Pesticides

Insecticides and Insect Repellents

Organophosphates penetrate the skin easily, however they have a very specific activity and have been shown to cause minimal skin irritation. Chlorpyrifos represent a popular group of organophosphates used in agriculture and household sprays. The data on the chlorpyrifos is not solid, however the 40.7% formula is thought to be a skin irritant. Naled (Dibrom) is labeled as a corrosive. Malathion has been shown to have a weak allergic contact sensitizer. However, of the cases reported by Schenker et al., there were 47 reports of urticaria, 38 reports of angioedema, and 213 reports of non-specific rash. Diazinon showed slight irritation in the Draize assay. In cases reported by Matsushita, diazinon was the organophosphate most associated with dermatitis. Parathion had a history of systemic poisoning in applicators and field workers.[12]

The carbamate insecticides have not been shown to cause irritation in the animal models. A few cases suggest that carbamates cause transient irritation, along with some reports of suspected allergic contact dermatitis. Willems described a case of hand eczema in a carnation grower.[13] The case was described as severe and it had an acute onset.

Among the organochlorines, DDT, with the half-life of 2-15 years, is no longer used in most countries. Abrams has noted that allergic contact dermatitis due to DDT is not a widespread problem. A compound structurally similar to DDT is dicofol (Kelthane), except dicofol contains a central hydroxyl group.

Biological Insecticides, Repellents, and Synthetic Materials

Pyrethrums and pyrethrins (Black Flag and Raid) are from extracts of the dried flower Chrysanthemum cinerariaefolium. Pyrethins are used in pet sprays, aerosols, and household sprays. Contact with pyrethins may cause mild erythema that can appear at 24 hours and substantially subside by 72 hours. Osimitz has summarized the dermatotoxicity of the compounds.[14]

Capsicum oleoresin-capsaicin is extracted from pepper plants and is used as insect and animal repellant. The product has been labeled as a sensitizer even though the Buehler sensitization assay showed a negative result.[15]

Fungicides and Antimicrobials

Fungicides function by inhibiting fungal sterol synthesis enzymes or affecting mitochondrial electron transport. The action that fungicides have against fungi on plants is the same interaction the fungicides may have with skin proteins leading to skin sensitization.

Phthalimido Compounds

Captan is a fungicide used on grapes, apples, almonds, and other crops as well as in treatment for pityriasis versicolor. Marzulli and Maibach demonstrated that captan in a concentration of 1% is a significant contact allergic sensitizer. Out of 205 human test subjects, 9 were sensitized.[16]

Except for the chlorinated carbon in the side chain, Captafol is nearly identical to Captan's structure. Captafol is also used as a fungicide and has been used in the prevention of blight on potatoes and fungal diseases on fruit. Captafol

is labeled as causing both delayed and immediate types of hypersensitivity along with irritant dermatitis. In a case report, Groundwater described the sudden appearance of wheezing, vesiculation, and edema of the face and handlers of welder that had contact with bags of captafol.[17]

Carbamates

Benomyl (Benlate) is a systemic carbamate fungicide that can function as a cholinesterase inhibitor and is used to control diseases of fruits, nuts, vegetables, and ornamental plants. In 1972, seven Japanese workers reported allergic contact dermatitis after working in a California greenhouse that had carnations sprayed with benomyl.

Thiocarbamates

Many plant diseases have been treated with Maneb (Dithane M-22) and Zineb (Lonacol). Matsushita tested maneb and zineb with the guinea pig maximization procedure and found both to be potent sensitizers with a high degree of mutual cross reactivity.[18]

Parasitic infestations in cats and dogs are treated with tetmosol. Tetmosol can also act as a fungicide. Cronin has reported a case of exacerbated dermatitis after contact with tetmosol.[19]

Copper Fungicides

Although copper fungicides have negative animal sensitization studies, copper fungicides are labeled as potential sensitizers since copper is identified as a human sensitizer. Most copper compounds reported cases are due to irritation. Hostynek has summarized copper dermatotoxicology.[20]

Amine, Nitro, and Nitrile Chlorinated Benzene Derivatives

Metalaxyl (Ridomil) is an analog of phenyalanine that inhibits fungal protein synthesis by interference with the synthesis of ribosomal RNA. Direct facial exposure to metalaxyl has resulted in facial swelling, photophobia, along with slight erythema.

CONCLUSION

Agricultural workers generally are medically underserved for economic and other reasons. These workers with dermatitis would benefit by a more efficient research data-base that would simplify and make more reliable the diagnosis of allergic and photoallergic contact dermatitis, as well as methods to decrease percutaneous penetration.

PATCH TEST CONCENTRATIONS

Proposed concentrations based on the literature (Table 1) should be viewed with caution. Abrams documents the softness of many of these concentrations. As most countries pesticide registrations do not require the this information, we generally lack full documentation to demonstrate that the concentration and solvent do not irritate in man sensitized volunteers, nor that the concentration is high to evidence a response in sensitized individuals. Several patch test supply companies have deleted these chemicals from their catalogues for a variety of reasons.

Abrams summarized the case report and experimental contact sensitization literature. The paucity of case reports may represent massive under represent massive under reporting, as (1) many agricultural workers have minimal access to medical and dermatologic care and (2) minimal access of standardized diagnostic patch test batteries. We await the intervention of interested parties (governments and unions) to mitigate the above.

TABLE 1. Reproduced by permission of Informa Healthcare, a division of Informa plc. Copyright 2001 from Pesticide Dermatoses, by Penagos H., O'Malley M. and Maibach H. (editors).[21]

Material Insecticides	CAS #	Concentration %	Vehicle	Ref. and Commentary
Organophosphates				
Acephate	30560-19-1	1	butanol	O'Malley et al., 1995
Azamethiphos		2	aqueous	Iliev and Elsner, 1997
Azinphos methyl	86-50-0	1	Vaseline	Jung et al., 1987; Lisi, 1992a
Bromophos	2104-96-3	0.1	Vaseline	Jung et al., 1987
Butonate	126-22-7	5	aqueous	Jung et al., 1987
Chlorpyrifos	2921-88-2	1	butanol	O'Malley et al., 1995
Diazinon	333-41-5	1	butanol	O'Malley et al., 1995
		1	petrolatum	Lisi, 1992a
Demephion	8065-62-1	0.5	aqueous	Jung et al., 1987, 1989
Dichlorvos (DDVP)	62-73-7	0.02	aqueous	Ueda et al., 1994
		1	petrolatum	Lisi, 1986; Lisi et al., 1986
		0.5	petrolatum	(Sharma and Kaur, 1990) 20 negative controls
		0.1,1	petrolatum	Mathias, 1983
		0.05	aqueous	Lisi, 1992a
		0.05-0.1	aqueous	Jung et al., 1987
Dichlorvos 50		1	aqueous	Stoermer, 1985
Dimethoate	60-51-5	1	petrolatum	Lisi, 1992a; Schena and Barba, 1992
		1	alcohol	Haenen et al., 1996
		1	Oil	Jung et al., 1989
		1-2	aqueous	Jung et al., 1987
Malathion	121-75-5	1	butanol	21 negative controls, O'Malley et al. 1995
		0.5	petrolatum	Guo et al., 1996; Jung et al., 1987; Lisi, 1992a; Sharma and Kaur 1990 Adams, 1990b; Cronin, 1980
		10	alcohol	Milby and Epstein, 1964
Methidathion	950-37-8	1-2	Vaseline	Jung et al., 1987
Mevinphos	7786-34-7	0.25	aqueous	
Naled	300-76-5	0.01	petrolatum	Guo et al., 1996
Naled technical		0.5	aqueous	Jung et al., 1987; + reactions noted in 4/6
		1	petrolatum	Lisi, 1992a
		60%	xylene	Edmundson and Davies, 1967

Material Insecticides	CAS #	Concentration %	Vehicle	Ref. and Commentary
Omethoate		1	aqueous	Haenen et al., 1996
Oxydemeton-methyl	301-12-2	1	Vaseline	Jung et al., 1987; Sharma and Kaur, 1990
Parathion ethyl	56-38-2	1	alcohol	Jung et al., 1987; Lisi 1986; Lisi, 1992a; Lisi et al., 1986
		1	petrolatum	van Ketel, 1976
		0.01	aqueous	Guo et al., 1996
Parathion methyl	298-00-0	1	petrolatum	Lisi, 1992a; Sharma and Kaur, 1990 irritant reactions in 2/20 controls; photopatch test, Jung et al., 1987; Lisi, 1986; Lisi et al., 1986
Phorate	298-02-2	1	petrolatum	Lisi, 1986; 1992a; Lisi, et al., 1986
Pirimiphos	29232-93-7	1	Vaseline	Jung et al., 1987
Quinalphos	13593-03-8	1	petrolatum	Sharma and Kaur, 1990
Trichlorfon	52-68-6	1	aqueous	Jung et al., 1987
N-methyl-carbamates				
Methomyl	16752-77-5	0.5	petrolatum	Ueda et al., 1994
		0.2 (a.i.); 1% formulated product	aqueous	Bruynzeel, 1991, 26 Controls
Carbaryl	63-25-2	1	petrolatum	Lisi, 1992a; Sharma and Kaur, 1990
Carbaryl 50%		0.25	Vaseline	Jung et al., 1989
Carbofuran	1563-66-2	1	petrolatum	Sharma and Kaur, 1990
Methiocarb	2032-65-7	0.5		Willems et al., 1997
Propoxur (Baygon EC 20)		1	aqueous	Stoermer, 1985
Organotin compounds				
Cyhexatin (plictran)		0.3	petrolatum	Ueda et al., 1994
Fenbutatin oxide	13356-08-6	0.1	butanol	O'Malley et al., 1995
Tributyltin oxide (TBTO)		0.01	aqueous	3/22 Controls had irritant at this concentration, Gammeltoft, 1978
Triphenyl tin fluoride		0.01,0.1,1,1	acetone	Andersen and Petri; 1982
Organochlorines				
Aldrin	309-00-2	1	petrolatum	Sharma and Kaur, 1990
DDT	50-29-3	1	petrolatum	Sharma and Kaur, 1990; Fisher, 1986
DDT		10%,l%	acetone	Cases of pemphigus negative on testing with 1:10 and 1: 100 dilutions Tsankov et al., 1998
DDT		1	petrolatum	Adams, 1990b
DDT technical		l	acetone, Vaseline	Jung et al., 1987
DDT lotion		5	acetone	Citing U.S. case reported by Leider; Behrbohm, 1962
DDT technical		undiluted	none	Behrbohm, 1962
Purified DDT		undiluted	none	Behrbohm, 1962
Endosulfan	115-29-7	0.5	petrolatum	Lisi 1986; Lisi et al., 1986
		1	petrolatum	Lisi, 1992a
Fentichlor	97-24-5	0.1,0.5,1.0	aqueous	Phototesting-persistent light reaction Norris et al., 1988
Lindane (Hexachlorocyclohexane)	58-89-9	1	petrolatum	Adams, 1990b; Cronin, 1980; Lisi, 1992a; Lisi et al., 1986; Sharma and Kaur, 1990
(Hexachlorocyclohexane technical)		not specified	acetone or Vaseline	Jung et al., 1987
(Hexachlorocyclohexane technical)		15-20	acetone	Behrbohm, 1962
Dienochlor		1	butanol	O'Malley et al., 1995
		1	petrolatum	10 negative controls; van Joost et al., 1983
Pyrethroid				
Allethrin		1	petrolatum	Lisi, 1992b
Cypermethrin		l	petrolatum	Lisi, 1992b; Sharma and Kaur, 1990
Deltamethrin		1	petrolatum	Lisi, 1992b

TABLE 1 (continued)

Material Insecticides	CAS #	Concentration %	Vehicle	Ref. and Commentary
Fenothrin		1	petrolatum	Lisi, 1992b
Fenvalerate		1	petrolatum	Sharma and Kaur, 1990
Permethrin		1	petrolatum	Lisi, 1992b
		1	butanol	O'Malley et al., 1995
Fluvalinate		1	butanol	O'Malley et al., 1995
Resmethrin		1	petrolatum	Lisi, 1992b
Pyrethrum		2	petrolatum	Dannaker et al., 1993; Jung et al., 1987; Lisi 1992a; Piraccini et al., 1991
Miscellaneous insecticides				
Arsenic trioxide		5	starch powder	
Benzyl benzoate	120-51-4	1	petrolatum	Stransky et al., 1996
Bacillus thuringiensis	68038-71-1			Adams
Chlorobenzene	108-90-7	5		Adams
DEET (diethyl toluamide)	134-62-3	5%	ethanol/ petrolatum	Lamberg and Muirennan, 1969
		0.1 cc undiluted Ready-to use product in open test; prick test	NA	Maibach and Johnson, 1975; Wantke et al., 1996
Imidacloprid		0.5	petrolatum	Willems et al., 1997
Nicotine		1	aqueous	Adams, 1990b
		5	aqueous	Lisi, 1992a
Propargite	2312-35-8	0.05		Irritant reactions in controls; Nishioka, et al., 1970
Oxythioquinox		0.1	petrolatum	Ueda et al., 1994
Piperonyl butoxide				Jung et al., 1987
Tetrachloroacetophenone		0.01	petrolatum	Lisi, 1992a
Insecticide precursors				
Ethoxymethylenemalono- nitrite		0.01-1	petrolatum	Wakelin et al., 1998
Malononitrile		0.01-1	petrolatum	Wakelin et al., 1998
TCAP (Tetrachloroacetophenone)		0.01-2	petrolatum	van Joost and Wiemer, 1991
Rodenticides				
Warfarin		0.05	petrolatum	Cronin, 1980
		0.05	Vaseline	Jung et al., 1987
Antimicrobials and antibiotics Isothiazolines				
Methylchloroisothiazolin one-methylisothiazolinone (Kathon CG)		0.01	aqueous	Zemtsov, 1991
Hexahydro, 1,3,5,tris (2- hydroxyethyl) triazine (Grotan)		0.2	aqueous	Keczkes and Brown, 1976
Sodium hypochlorite	10022-70-5	1	petrolatum	Hostynek et al., 1990
		1	aqueous	Urticarial reaction in open test; Hostynek et al., 1989
Tetrachlorsalicylanilide	1154-59-2	0.1	alcohol	Wilkinson, 1961
Piscicide				
Rotenone		5%	talcum	[Lisi, 1992 #833; Fisher, 1986 #1965]
		1%	talcum	[Adams, 1990 #1992]
Fungicides **Phthalimido compounds**				
Captafol, Difolatan	2939-80-2	0.05-0.1, 0.1	petrolatum	Lisi, 1986; Lisi et al., 1986
		1	petrolatum	Camarasa, 1975; Groundwater, 1977; Sharma and Kaur, 1990; 2 irritant reactions in 20 controls

Material Insecticides	CAS #	Concentration %	Vehicle	Ref. and Commentary
Captafol, Difolatan		0.1	petrolatum	Guo et al., 1996; Peluso et al., 1991; Piraccini et al., 1991
		1,5,10	aqueous	Brown, 1984
Captan	133-06-2	0.25	petrolatum	Jung et al., 1987 Adams, 2
Captan		0.1	petrolatum	Dannaker et al., 1993; Piraccini et al., 1991, concentrations used in material from Troh Lab/Pharmascience
		1	petrolatum	Rudner, 1977
Folpet	133-07-3	0.05		Adams, 2
		0.1	petrolatum	Guo et al., 1996; Lisi 1992a; Peluso et al., 1991; Piraccini et al., 1991
Ditalimfos (plondrel)	5131-24-8	0.01	petrolatum	Jung et al., 1987; van Ketel 1975; van Ketel 1977
Phaltan	133-07-3	0.01	petrolatum	Adams, 1990 #1992]
Carbamates and carbamate-like compounds				
Benomyl	17804-35-2	0.1	petrolatum or water	Dannaker et al., 1993; Jung et al., 1987; Kuhne et al., 1985; Piraccini et al., 1991 Lisi, 1986
Benomyl	17804-35-2	1.0	petrolatum	No reaction in 10 controls; Ueda et al., 1994; van Joost et al., 1983; van Ketel, 1976
		undiluted and diluted 1:5	olive oil	Savitt, 1972
Thiocarbamates				
Mancozeb	8018-01-7	1	petrolatum	Lisi, 1992a
Mancozeb/pyrifenox		2	aqueous	Iliev and Elsner, 1997
Maneb	12427-38-2	1	petrolatum	Lisi, 1992a; Peluso et al., 1991; Piraccini et al., 1991; Sharma and Kaur, 1990; Adams and Manchester, 1982; Nater et al., 1979
		1	petrolatum	Dannaker et al., 1993
Maneb technical		0.5	vaseline	Jung et al., 1989; Jung et al., 1987
		1	aqueous	Jung et al., 1987
Zineb	12122-67-7	0.002%	not specified	low concentration selected as 1 % of end use concentration–testing positive in 2 patients; negative in 3 controls; Kleibl and Rackova, 1980
		1	petrolatum	Adams, 1990a; Dannaker et al., 1993; Jung et al., 1987; Lisi, 1992a; Piraccini et al., 1991
Metiram - ammonia complex of zinc EBDC	9006-42-2	1	petrolatum	Koch, 1996
Thiram (TMTD)	137-26-8	1	petrolatum	Lisi, 1992a; Rudzki and Napiorkowska, 1980; Sharma and Kaur, 1990; Adams, 2
		2	Vaseline	Jung et al., 1987
		1	petrolatum	Cronin, 1980
Disulfiram (TETD)	European standard		petrolatum	
Tetmosol (sulfiram)	95-05-6	1	petrolatum	Cronin
Propineb	12071-83-9 9016-72-2	0.01-1	aqueous	Jung et al., 1989
Ziram	137-30-4	1	petrolatum	Dannaker et al., 1993; Guo et al., 1996; Jung et al., 1987; Lisi, 1992a
		0.1	petrolatum	Piraccini et al., 1991

TABLE 1 (continued)

Material Insecticides	CAS #	Concentration %	Vehicle	Ref. and Commentary
Nitro and nitrile chlorinated benzenes				
Chlorothalonil	1897-45-6	0.001	petrolatum or acetone or butanol	Flannigan et al., 1986; O'Malley et al., 1995
		0.005 0.0001	aqueous	Liden, 1990; 10 controls negative at 0.0001%
		0.02	aqueous	Ueda et al., 1994
		0.01	aqueous	Immediate anaphylactoid reaction observed on open test, Dannaker et al., 1993
		0.01	Pet	Lisi, 1992a
		0.01	acetone	No reaction in 10 individuals; Bach and Pedersen, 1980; no reactions in unspecified number of controls; Johnsson et al., 1983
		0.002	aqueous	Patch and photopatch–stronger reaction on; Matsushita et al., 1996
		0.01	petrolatum	Borderline toxic reactions at 48-72 h in 4/17 control subjects; Bruynzeel and van Ketel, 1986
Tetrafluoro-terephthalonitrile	1835-49-0	0.001-0.1	petrolatum	Carmichael et al., 1989
PCNB (pentachloronitro-benzene)	82-68-8	0.5		Cronin
		10	aqueous	Jung et al., 1987
		1	petrolatum	Sharma and Kaur, 1990
		1	butanol	O'Malley et al., 1995
Dichloronitroaniline		1	butanol	O'Malley et al., 1995
Dinobuton	973-21-7	1	petrolatum	Adams, 2 Cronin
		30-40	not specified	Wahlberg, 1974
Dinocap	39300-45-3	0.5	petrolatum	Sharma and Kaur, 1990
		1	petrolatum	Lisi, 1992a
DNCB		0.01		Cronin
		0.25	aqueuos	Malten, 1974
Rodannitrobenzene		1	petrolatum	Fregert, 1967
Dichlomitroaniline		1	Butanol	O'Malley et al., 1995
Trichlorodinitrobenzene		0.01	aqueous	Jung et al., 1987
Miscellaneous fungicides				
Antisapstain		1		Adams, 2
Benzimidazole carbamate		1	petrolatum	Sharma and Kaur, 1990
Carbendazim		0.5	petrolatum	Bruynzeel et al., 1995
$CuSO_4$	7758-98-777	1	petrolatum	Lisi, 1986 ; Lisi 1992a; b; Lisi et al., 1986; 1987
Copper fungicide (uns)		1	petrolatum	Sharma and Kaur, 1990
Dichlofluanid	1085-98-9	10%	petrolatum	Hansson and Wallengren, 1995; testing on controls not described
Dichlorophen		0.5		Adams, 2
Dithianon	3347-22-6	1	petrolatum	Calnan, 1969; Jung et al., 1987; Koch, 1996; Lisi 1992a; van Joost et al., 1983
Dodemorph	1593-77-7	0.25	petrolatum	van Ketel, 1975
Fluazinam	79622-59-6	0.5	petrolatum	Bruynzeel et al., 1995
Iprodione	36734-19-7	1		default
Ketoconazole	65277-42-1	1	ethanol	Valsecchi et al., 1993
Methyl-2-benzimidazole		1	petrolatum	Sharma and Kaur, 1990
Pyrifenox (as part of Rondo-M)	88283-41-4	2	aqueous	Iliev and Elsner, 1997

Material Insecticides	CAS #	Concentration %	Vehicle	Ref. and Commentary
Triforine (Saprol)		0.02	aqueous	58 male subjects and 47 female in cross sectional study; Ueda et al., 1994. No control subjects identified–irritant threshold 0.5% comparable to dichlorvos 1% in guinea pig studies; strong allergic reaction on guinea pig test
Streptocycline		3	petrolatum	Sharma and Kaur, 1990
Streptomycin		1		Adams, 2 citing van Ketel, 1974 - citation not located in bibliography
Sulfur	7704-34-9	1	butanol	O'Malley et al., 1995
Inorganic sulfur	63705-05-5	5	petrolatum	Sharma and Kaur, 1990
		5-20	alcohol, Vaseline	Schneider, 1978
		1,5	petrolatum	Wilkinson, 1975
Triadimefon	43121-43-3	1	petrolatum	UC Davis Unpublished study
Vinclozolin	50471-44-8	1	butanol	O'Malley et al., 1995
Dichlone	117-80-6	NS	not specified	Adams, 2
Animal feed additives				
Olaquindox	23696-28-8	0.01-5.0	petrolatum	Schauder et al., 1996
Ethoxyquin	91-53-2	0.5	petrolatum	Savini et al., 1989
Fumigants				
Chloropicrin		0.25	petrolatum	Adams, 2
Dazomet		0.1	petrolatum	Jung et al., 1987; Vilaplana et al. 1993, Black, 1973
		0.1-0.25	petrolatum	Lisi et al., 1986
		0.025	Vaseline	Warin, 1992; negative tests in 20 controls
		0.0025		
Metam sodium		0.03	aqueous/Vaseline	Jung et al., 1987
		0.1	aqueous	Schubert, 1978
		0.5-1	petrolatum	Lisi et al., 1986
MITC		0.1	aqueous	Schubert, 1978
1,2 dichloropropane		10%	petrolatum	Baruffini et al., 1989; material used as a solvent in painting, metalworking
		1%	petrolatum	2 patients, no controls tested; Kleibl and Rackova, 1980
D-D mixture(dichloropropene and dichloropropane)	26952-23-8	0.05	petrolatum	Bousema et al., 1991; 20 negative controls; patient was positive at concentrations ranging from 0.005 2%
		0.02-2.0	petrolatum	van Joost and de Jong, 1988
Herbicides Phenoxy compounds				
2,4-D	94-75-7	1	petrolatum	Sharma and Kaur, 1990
		1-2	aqueous	Jung et al., 1987 Adams, 2
2,4-D/2,4,5-T		0.4	diesel oil	Jung, 1975 #1237
2,4-D/2,4,5-T		0.1-2	olive oil	
p-chloro-o-cresol		1	aqueous	Fregert, 1968
MCPA		5-10	aqueous	Jung et al., 1987
		1	aqueous	Jung et al., 1989
2,4,5-T				Tested as part of a mixture with 2,4 D; Jung et al., 1987
Dichlorprop		1	aqueous	Jung et al., 1989
Dicamba		0.5	aqueous	Jung et al., 1989
Acetanilides				
Alachlor (Lasso)	15972-60-8	0.1	petrolatum	Vilaplanaetal., 1993; Adams, 2 citing Iden and Schroeter, 1977
		1	petrolatum	Lisi et al., 1986
Allidochlor (Randox)	93-71-0	0.01-0.1	petrolatum	Cronin, 1980
Butachlor		1	aqueous	Sharma and Kaur, 1990

TABLE 1 (continued)

Material Insecticides	CAS #	Concentration %	Vehicle	Ref. and Commentary
Propachlor (Ramrod)		1	?	Schubert, 1979
		0.05	aqueous	Jung et al., 1987 citing Korossy 1978 meeting presentation
Propachlor		0.01-0.5	aqueous	Jung et al., 1987; 1989
Triazines				
Atrazine	1912-24-9	0.1		Jung et al., 1987; Vilaplana et al., 1993
		1	petrolatum	Sharma and Kaur, 1990
Amitrole		1		English et al., 1986
Simazine		NS	not specified	Jung et al., 1987
		5-10	aqueous	Jung et al., 1987
Ametryn		0.1	aqueous	Jung et al., 1987
Carbamates				
Carbyne (Barban)	101-27-9	10	mineral oil	Brancaccio and Chamales 1977
Molinate	2212-67-1	1	petrolatum	Lisi et al., 1986
Thiobencarb		1	aqueous	Sharma and Kaur, 1990; 2 irritant reactions
Chloropropham		0.5	Vaseline	Jung et al., 1987
Bipyridyls				
Diquat		0.1	petrolatum	Lisi et al., 1986; Vilaplana et al., 1993
Paraquat		0.1	petrolatum	Lisi et al., 1986; Sharma and Kaur 1990; Vilaplana et al., 1993
Nitrobenzene				
4,6-DiNitroOrthoCresol	534-52-1	0.5-1	aqueous	Lisi et al., 1986
Dinitro-o-cresol		0.5		Lisi, 1986; Lisi et al., 1986
		50	aqueous	Jung et al., 1987
Dinoterbe				Sabouraud et al., 1997
Nitralen	4726-14-1	1	?	Nishioka et al., 1983
Nitrofen		0.5	olive oil	Jung et al., 1987
Chloranil	118-75-2	1		Adams, 2
Chloridazon	1698-60-8	0.1		Lisi et al., 1986
Miscellaneous				
Anilofos		I	aqueous	Sharma and Kaur, 1990
Cacodylic acid		0.1-1%	aqueous	Bourrain et al., 1998
Cymoxanil	57966-95-7			
Dichlofopmethyl		1	aqueous	Sharma and Kaur, 1990
Fluchloralin		1	aqueous	Sharma and Kaur, 1990
Glyphosate	1071-83-6	10	aqueous	Maibach, 1986
Glyphosate end use formulation with unspecified concentration of active ingredient		1-5	aqueous	Recorded reaction to preservative ingredient, Hindson and Diffey1984a; 1984b
Hexythiazox		1	petrolatum	Willems et al., 1997
Oxadiazon		1	aqueous	Sharma and Kaur, 1990
Pendimethalin		1	aqueous	Sharma and Kaur, 1990
Phenmedipham	13684-63-4	2		Adams, 2
Propanil	709-98-8	1	petrolatum	Lisi et al., 1986
		0.1	petrolatum	Vilaplana et al., 1993
Trichlorobenzyl chloride	1344-32-7	1		Adams, I
Choline chloride (chlormequat chloride) (PGR)	999-81-5	1		Adams, 2
Dodecyl gallate		0.25	petrolatum	Guo et al., 1996
Chloropropham	101-21-3	0.5	Vaseline	Jung et al., 1987
Dalapon	75-99-0	50	aqueous, Vaseline	Jung et al., 1987
Ethephon	16672-87-0	2	aqueous	Jung et al., 1987
Methabenzthiazinon		1	petrolatum	Sharma and Kaur, 1990
Metoxuron		1	petrolatum	Sharma and Kaur, 1990

Material Insecticides	CAS #	Concentration %	Vehicle	Ref. and Commentary
Isoproturon		1	petrolatum	Sharma and Kaur, 1990; 2 irritant reactions
Nitrofurazone		l	petrolatum	Guo et al., 1996
Fenazox	495-48-7	unknown	not specified	Jung et al., 1987

REFERENCES

1. Chew A-L, Maibach H, editors. Irritant Dermatitis. Springer; 2006.

2. Menne T, Maibach H. Exogenous Dermatoses: Environmental Dermatitis. Boca Raton (FL): CRC Press; 1991.

3. Maibach HI. Irritation, sensitization, photoirritation and photosensitization assays with a glyphosate herbicide. Contact Dermatitis. 1986 Sep;15(3):152-6.

4. Maibach HI and Johnson HL. Contact urticaria syndrome. Arch. Dermatol. 1975 Apr;111(6):726-30.

5. Villarama C, Maibach H. Correlations of patch test reactivity and repeated open application test (ROAT)/provocative use test (PUT). Food Chem Toxicol. 2004 Nov;42(11):1719-25.

6. Frosch P, Schulze-Dirks A, Hoffmann M, Axthelm I, Kurte A. Efficacy of skin barrier creams (I). The repetitive irritation test (RIT) in the guinea pig, Contact Dermatitis. 1993 Feb;28(2):94-100.

7. Marzulli F, Maibach H. The rabbit as a model for evaluating skin irritants: a comparison of results obtained on animals and man using repeated skin exposures. Food Cosmet Toxicol. 1975 Oct;13(5):533-40.

8. Uttley M, Van Abbe N. Primary irritation of the skin: mouse ear test and human patch test procedures. J Soc Cosmet Chem. 1973;24:217-27.

9. Patrick E, Maibach H. A novel predictive assay in mice. Toxicologist. 1987;7:84.

10. Lanman B, Elvers W, Howard C. The role of human patch testing in product development program. In: Proceedings, Joint Conference Cosmetic Sciences, Toilet Goods Association, Washington, DC, 1968, 135-145.

11. Phillips L, Steinberg M, Maibach H, Akers W. A comparison of rabbit and human skin responses to certain irritants. Toxicol Appl Pharmacol. 1972;21:369-82.

12. Matsushita T, Aoyama K, Yoshimi K, Fujita Y, Ueda A. Allergic contact dermatitis from organophosphorus insecticides. Ind. Health. 1985;23(2):145-53.

13. Willems PW, Geursen-Reitsma AM, van Joost T. Allergic contact dermatitis due to methiocarb. Contact Dermatitis. 1997 May;36(5):270.

14. Franzosa JA, Oximitz TG, Maibach HI. Cutaneous contact urticaria to pyrethrum–real?, common?, or not documented?: an evidence-based approach. Cutan Ocul Toxicol. 2007;26(1):57-72.

15. California Department of Pesticide Regulation, Product Label Database-ready to use deer and rabbit repellent containing 0.216% of allyl isothiocyanate and 0.625% of capsicum oleoresin; 1998.

16. Marzulli F, Maibach H. Antimicrobials: experimental contact sensitization in man. J Soc Cosmet Chem. 1973;20:67.

17. Groundwater JR. Difolatan dermatitis in a welder; non-agricultural exposure. Contact Dermatitis. 1977 Apr; 3(2):104.

18. Matsushita T, Arimatsu Y, Nomura S. Experimental study on contact dermatitis caused by dithiocarbamates maneb, mancozeb, zineb, and their related compounds. Int Arch Occup Environ Health. 1976 Jul; 37(3):169-78.

19. Cronin E. Pesticides. In: Contact dermatitis. Edinburgh:Churchill Livingstone; 1980. 393 p.

20. Hostynek J, Maibach H. Copper and the Skin. New York: Informa Healthcare; 2006.

21. Penagos H, O'Malley M, Maibach H. Pesticide Dermatoses, Dermatology: Clinical and Basic Science Series. Informa Healthcare; 2001.

doi:10.1300/J096v12n01_02

Neurotoxicity of Pesticides

Matthew C. Keifer, MD, MPH
Jordan Firestone, MD, MPH, PhD

SUMMARY. Several pesticides such as organophosphates, carbamates and the organochlorine pesticides directly target nervous tissue as their mechanism of toxicity. In several others, such as the fumigants, the nervous system is affected by toxicological mechanisms that diffusely affect most or all tissues in the body. Both the central and peripheral nervous system are involved in the acute toxidromes of many pesticides resulting in acute short-term effects. There is strong human epidemiological evidence for persistent nervous system damage following acute intoxication with several important pesticide groups such as organophosphates and certain fumigants. However, whether persistent nervous system damage follows chronic low-level exposure to pesticides in adults (particularly organophosphpates), and whether in utero and/or early childhood exposure leads to persistent nervous system damage, is a subject of study at present. Parkinson's Disease, one of the most common chronic central nervous system diseases, has been linked to pesticide exposure in some studies, but other studies have failed to find an association. Several new pesticidal chemicals such as the neo-nicotinoids and fipronil have central nervous system effects, but only case reports are available to date on acute human intoxications with several of these. Little data are yet available on whether long-term effects result from these chemicals. Several ongoing or recently completed studies should add valuable insight into the effects of pesticides on the human nervous system particularly the effect of low-dose, chronic exposure both in adults and children. doi:10.1300/J096v12n01_03 *[Article copies available for a fee from The Haworth Document Delivery Service: 1-800-HAWORTH. E-mail address: <docdelivery@haworthpress.com> Website: <http://www.HaworthPress. com> © 2007 by The Haworth Press, Inc. All rights reserved.]*

KEYWORDS. Pesticides, insecticides, fumigants, toxicity, nervous system diseases/chemically induced, Parkinson's Disease, peripheral neuropathy

INTRODUCTION

Pesticides can damage the nervous system in various ways. Pesticides used to kill higher order pests, such as insects, often target the nervous system directly. Other pesticides produce adverse neurological effects by disrupting general cellular mechanisms necessary to support the high metabolic activity of the nervous system. In this paper, we discuss those pesti-

Matthew C. Keifer is Associate Professor, University of Washington, Departments of Medicine and Environmental & Occupational Health Sciences, and Jordan Firestone is Acting Assistant Professor of Neurology, University of Washington UW Medicine.

Address correspondence to: Matthew C. Keifer, MD, MPH, University of Washington, Department of Environmental and Occupational Health Sciences, Box 357234, Seattle, WA 98195-7234 (E-mail: mkeifer@u.washington. edu).

[Haworth co-indexing entry note]: "Neurotoxicity of Pesticides." Keifer, Matthew C., and Jordan Firestone. Co-published simultaneously in *Journal of Agromedicine* (The Haworth Medical Press, an imprint of The Haworth Press, Inc.) Vol. 12, No. 1, 2007, pp. 17-25; and: *Proceedings from the Medical Workshop on Pesticide-Related Illnesses from the International Conference on Pesticide Exposure and Health* (ed: Ana Maria Osorio, and Lynn R. Goldman) The Haworth Medical Press, an imprint of The Haworth Press, Inc., 2007, pp. 17-25. Single or multiple copies of this article are available for a fee from The Haworth Document Delivery Service [1-800-HAWORTH, 9:00 a.m. - 5:00 p.m. (EST). E-mail address: docdelivery@haworthpress.com].

Available online at http://ja.haworthpress.com
© 2007 by The Haworth Press, Inc. All rights reserved.
doi:10.1300/J096v12n01_03

cides recognized as the most important neurotoxins.

The paper is divided into sections on central and peripheral nervous system effects. Within each section, we describe the acute and chronic neurological effects of each chemical or group of chemicals, and we identify the specific target within the nervous system and the mechanism of toxicity where known. For example, several different pesticide groups exert their effects by altering the function of ion channels. The toxic mechanisms involve increased sodium influx through voltage-gated sodium channels or decreased chloride efflux through ligand-gated (GABA) chloride channels. In either case, the chemical foot in the ionic door depolarizes neuronal membranes, resulting in neuronal hyper-excitability.

While many neurotoxic pesticides work through multiple mechanisms, affecting either the central or peripheral nervous system, or both, we have focused our discussion on each pesticide's most notorious or well-described toxic effect. Of course, some pesticides merit discussion for more than one reason, so may be mentioned in more than one context. Throughout the paper, the terms acute and chronic will be used to distinguish between a pesticide's immediate or early neurotoxic effects (acute) and its residual or prolonged neurotoxic effects (chronic). Short-term and chronic will be used in reference to the duration of pesticide exposure.

CENTRAL NERVOUS SYSTEM

Symptoms of central nervous system dysfunction are among the most dramatic events in clinical medicine. Acute symptoms can include headache, nausea, dizziness, sensory paresthesias, incoordination, and movement disorders with dystonia or tremor. Toxicity often involves neuronal hyperexcitability, producing myoclonic jerking or generalized tonic-clonic seizures. Disorders of cognition with confusion or loss of consciousness may also occur. Chronic weakness may include myelopathic features (e.g., spasticity).

Several groups of pesticides can affect the central nervous system. Among them are the organochlorines, pyrethrins, organophosphates, fumigants, and some novel agents.

Organochlorines

The organochlorine pesticides are an excellent example of toxins that induce ion channel dysfunction resulting in central neurological effects. The most famous of the organochlorines, DDT, affects sodium channel function by preventing the decrease in sodium permeability that normally follows an action potential, thereby facilitating repetitive neuronal discharges. The organochlorines are used less today than they were in the past, because DDT and many other organochlorines lost their U.S. registration, principally due to concerns about environmental persistence. Nonetheless, DDT is still manufactured for use in other countries, and endosulfan, a less environmentally persistent organochlorine, is still registered for use in the U.S.

Cyclodienes (which include endosulfan, heptachlor, aldrin, dieldrin, and chlordane) and the gamma isomer of hexachlorohexane, known as lindane, interact with GABA receptors to alter chloride currents, resulting in neuronal excitation. Neurological effects are often the initial presentation of cyclodiene exposure, and although symptoms may be delayed, once begun they may last many hours.[1,2]

One organochlorine known as chlordecone (kepone) was responsible for an epidemic of poisonings at a manufacturing plant in Hopewell, Virginia, in the late 1970s. These cases included workers as well as some of their spouses. Neurological symptoms included nervousness, tremor, and opsoclonus.[3] The toxicological mechanism of chlordecone is similar to that of the cyclodienes. The half-life of the chemical in blood is in the order of 165 days, but this can be foreshortened with the used of cholestyramine. Subsequent evaluation six years after initial diagnosis suggested that symptoms abated with reduction of blood levels of the pesticide.[4]

Hexachlorophene, an antiseptic used in many soap products, is a potent central nervous system neurotoxicant. Toxicity can occur either acutely or through chronic exposure, and from dermal or oral absorption. Neurological effects include cerebral edema and spongy degeneration of white matter. Optic nerve atrophy and blindness have also been reported.[5,6]

Pyrethrins

The pyrethrin insecticides are derived from the chrysanthemum flower. Pyrethrum is a mixture of six natural pyrethrins, and the pyrethroids are semisynthetic analogs with greater environmental stability and lower antigenicity. Pyrethroids are divided into two groups based on their neurological effects in high dose rodent studies: the C-S group (choreoathetosis-salivation-seizures) and the T group (tremors). These compounds alter neurological function by interacting with sodium and chloride channels.[7] High doses are rarely seen in human exposures, so neurological effects are not commonly seen in humans.[1] Nonetheless, the pyrethroids have become a very important insecticide group, occupying second place to the organophospates in terms of insecticide dollars spent per year.[2]

Cholinesterase Inhibitors

The cholinesterase inhibitors are a group of insecticidal chemicals that includes some of the most toxic chemicals known to man. Among these are the organophosphates and the N-methyl-carbamates. These chemicals induce acute neurological effects by inhibiting the cholinesterase enzyme. Cholinesterase provides an essential function in cholinergic neurotransmission by degrading acetylcholine in the synaptic cleft. This enzyme activity is essential for normal cholinergic function throughout the central nervous system and the sympathetic, parasympathetic, and motor components of the peripheral nervous system. If the enzyme is inhibited, often permanently by organophosphates or temporarily by N-methyl-carbamates, acetylcholine accumulates in the synaptic cleft, leading to over-stimulation of glands, nerves, and muscles.

The organophosphates include parathion, methamidiphos, azinphos methyl, chlorpyrifos, and phosdrin. These chemicals bind to the cholinesterase enzyme, which can become irreversibly inactivated as an "aged" enzyme-pesticide covalent complex. This aging process can be prevented through competition with the oxime compounds, such as pralidoxime if

given early in the intoxication. Because the N-methyl-carbamates, such as aldicarb, carbaryl, and methomyl, bind reversibly to the cholinesterase enzyme, aging does not occur, and chronic neurological effects are less common. For this reason the toxic effects of the N-methyl-carbamates are most evident while elevated levels of active pesticides are present.

Acute intoxication with organophosphates or carbamates includes symptoms related to parasympathetic over-stimulation, including salivation, lacrimation, urinary incontinence, diarrhea, gastroenteric cramping, and emesis (remembered by the acronym SLUDGE). Acute central neurological symptoms may include confusion, loss of consciousness, depressed respiratory drive, and convulsions. Acute peripheral neurological manifestations may affect motor function, with fasciculations, weakness, loss of muscle control, and respiratory paralysis. The combination of excessive secretions, brochospasm, decreased respiratory drive, and weakened muscular tone may lead to hypoventilation and suffocation.[2]

Persistent central nervous system effects have also been documented in subjects who have suffered an acute organophosphate poisoning.[8-13] One investigator has suggested that intoxication with N-methyl carbamates can produce similar decrements of neurobehavioral function.[9] The cognitive changes in these cases have been documented through the use of sensitive neurobehavioral testing, revealing subtle changes in brain function. Some studies have also suggested that similar, persistent neurological changes may result from repeated exposure to organophosphates at levels below the threshold for acute intoxication.[14-16] However, this finding has been inconsistent.[17,18]

Fumigants

Fumigants are notorious for their serious acute and chronic central nervous system effects, so many of these chemicals are no longer used. Among those with important central neurological effects that are still registered for use are methyl bromide, ethylene oxide and sulfuryl fluoride.

Methyl bromide has been responsible for many acute intoxications and deaths. Both pulmonary and central nervous system toxicity appear to play a part.[19,20] Adverse effects have also been reported from persistent exposure, without episodes of acute intoxication.[21,22] Chronic effects to include both the central and peripheral nervous system.[23,24] The mechanism of toxicity is not well elucidated, although methyl bromide's alkylating capability may be involved.[23]

Ethylene oxide is an agent used to sterilize heat-sensitive materials. Chronic effects involving both the central and peripheral nervous system have been reported following acute, high-level exposure. Some reports suggest that both short-term high level and chronic low level exposures may result in chronic central neurological effects.[25,26]

Sulfuryl fluoride has been less frequently reported as an acute central nervous system intoxicant.[27,28] However, two epidemiological studies have demonstrated chronic central nervous system deficits associated with chronic exposure to sulfuryl fluoride in pesticide applicators.[29,30]

Novel Agents

Several newer insecticidal groups have shown acute central nervous system effects. Amitraz, an insecticide and miticide, belongs to the amidine chemical family. It is an alpha-2 receptor agonist and has been the subject of several clinical toxicity series reported from Turkey. Children with either dermal or gastrointestinal exposures to the chemical presented with dizziness. Cases with more severe central nervous system depression were also reported to include somnolence, coma, convulsions, and respiratory failure requiring mechanical ventilation. Autonomic effects including hypotension or hyptertension, bradycardia, fever, hyperglycemia, and miosis were sometimes seen. Onset generally occurred promptly after exposure. Fortunately, the prognosis for recovery appears to be good.[31-33]

Fipronil, an insecticide widely used for structural and veterinary applications, also acts by binding to GABA receptors and increasing neuronal excitability.[34] Human toxicity data is very limited regarding this relatively new chemical, but the presenting characteristics of seven cases were summarized by Muhamed et al.[35] The toxidrome (i.e., the constellation of signs and symptoms that suggest a specific class of poisoning) included vomiting, agitation, and convulsions. Medical records review identified three patients with tonic clonic seizures. Two patients had non-sustained convulsions, and one died from continuous, uncontrollable seizures.

Parkinson's Disease

Parkinson's Disease (PD) is an extrapyramidal movement disorder whose etiology in most cases is uncertain. However, there is mounting evidence that pesticide exposure may be an important environmental risk factor for PD.[36] Toxicological studies *in vitro*[37,38] and *in vivo*[39,40] have demonstrated specific neurodegenerative effects from exposure to certain pesticides, and human case reports have suggested a causal relationship between certain pesticide exposures and PD.[41,42] Although epidemiological evidence linking PD with pesticides has been inconsistent, some studies have detected increased risk from surrogates for pesticide exposures, such as well water consumption or living in rural, agricultural regions.[43-47] More consistent results have been seen with occupational exposures such as with farming activities or direct exposure to herbicides or insecticides.[47-51] A methodological limitation of many studies has been the difficulty in identifying a specific chemical. Instead, associations with classes of pesticides are commonly reported.

The Agricultural Health Study (AHS) is a large-scale, prospective epidemiological study that is particularly relevant to the question of pesticide effects on the nervous system. Recent results from the AHS suggest that among 18,782 pesticide applicator subjects, those who used more insecticides were more likely to report neurological symptoms.[52] There was a slight association with fungicides as well, but the strongest effect was seen with organophosphate and organochlorine pesticides. This effect was independent of a history of acute pesticide poisoning. The AHS is an ongoing study, and more results are anticipated in the coming years.

PERIPHERAL NERVOUS SYSTEM

The peripheral nervous system is composed of the sensory, voluntary motor, and autonomic nervous systems. Any of these components can be damaged by pesticide exposures, although particular exposures have a predilection for particular components. As with central neurological effects, peripheral effects may be acute or chronic.

Pyrethrins

The pyrethrins and pyrethroids act on the peripheral nervous system through their interaction with sodium channels. These exposures manifest principally through their effect on the sensory nervous system. People with dermal exposure report numbness of the extremities and mucous membranes, particularly the facial and perioral areas. This is likely due to repetitive neural discharges triggering sensory perceptions. To date, no chronic peripheral nervous system effects have been associated with pyrethroids.

Organophosphates

One of the most intriguing presentations of chronic pesticide related illness is organophosphate induced delayed polyneuropathy (OPIDP). This predominantly motor neuropathy can occur with only a handful of organophosphates.[53,54] An organophosphate's ability to induce OPIDP appears to depend on its ability to inhibit a neuron membrane-associated protein known as neuropathy target esterase.[55,56] OPIDP is usually heralded by the onset of pain and cramping of the lower extremities beginning around two weeks after acute organophosphate intoxication. Sensory changes are generally minor, but motor dysfunction may be severe, characterized by a progressive, ascending, flaccid paralysis which may be only partially reversible. Cases of OPIDP following triortho-cresyl-phosphate poisoning suggest that concomitant, myelopathic effects with spastic weakness may complicate assessment of the most severely affected individuals.

Also associated with severe organophosphate intoxication is a poorly understood condition known as the intermediate syndrome. This syndrome begins one to several days after acute intoxication and is characterized by proximal muscle weakness. At times this can be severe enough to cause respiratory failure. Fortunately, the condition appears to be somewhat uncommon and to affect only those patients with severe, acute intoxications. Although the exact cause is unclear, pathology specimens have revealed degeneration of the motor endplate at the neuromuscular junction, suggesting injury related to sustained depolarization. Pralidoxime may help prevent this syndrome, and inadequate use of pralidoxime has been implicated in the intermediate syndrome.[57] However, OPIDP has been reported despite aggressive use of pralidoxime.[58] Clinicians should be aware of this syndrome, as patients in whom treatment with atropine has adequately controlled the early cholinergic crisis (SLUDGE: salivation, lacrimation, urination, diarrhea, gastroenteric cramping, and emesis) may nonetheless quietly drift off into respiratory failure if not closely observed.

Neo-Nicotinoids

Neo-nicotinoids are new insecticidal chemicals that selectively activate nicotinic receptors. In humans, nicotinic receptors are abundant in smooth muscle cells and are the predominant receptor of the postganglionic sympathetic nervous system. The structural differences between insect and mammalian nicotinic receptors confer selective toxicity for insects,[59] so the risk to humans may be low. To date, only a single clinical case report of human toxicity from neo-nicotinoids has appeared in the medical literature.[60] This case report was based on ingestion of a mixture of 9.7% imidacloprid, surfactant, and N-methyl-pyrrolidone. The patient presented with dizziness, disorientation, drowsiness, gastroesophageal erosions, cough, and fever, along with hyperglycemia and leukocytosis. Which of these symptoms was a direct result of nicotinic receptor activation is unclear.

Herbicides

Mixed sensory-motor peripheral neuropathies have been reported following high dose acute exposure to chlorphenoxy herbicides, such as

2,4-D and 3-T,C, through either ingestion or dermal exposure. It appears that these chemicals exert toxicity through uncoupling oxidative phosphorylation, damaging cell membranes and interfering with acetylcoenzyme A metabolism.[61] Whether the ultimate damage is caused by these chemicals themselves, or a common contaminant of older chlorphenoxy heribicides, dioxin, is unclear.

Fumigants

Among the fumigants, both methyl bromide and ethylene oxide have been associated with peripheral neuropathy. Several cases have been reported following acute overexposure to methyl bromide. These are often accompanied by substantial central nervous system damage, and recovery may be slow and incomplete.[62,63] Ethylene oxide also appears capable of inducing peripheral and central nervous system toxicity, even with low level, chronic exposure. Sural nerve biopsy has demonstrated axonal injury in several cases.[28,64]

Metals

Thallium, as a sulfate salt, was used as a rodenticide until 1972. It is no longer used for this purpose in the U.S., but poisonings are periodically reported as attempted homicides. In its ionized form, thallium acts like potassium, gaining access to the intracellular environment, where it interferes with multiple cellular processes. Though it is known to be toxic to many cell types, the exact mechanism of thallium toxicity is not clear. Its toxicity is manifest in both the central and peripheral nervous system, and acute and chronic peripheral neuropathy are common.[65,66] One recognizable and somewhat unique characteristic of thallium intoxication is alopecia.

Arsenic is another metal that has been used as a pesticide. In the past it was used as a rodenticide and, in the form of lead-arsenate, as a fungicide for crops. It is used in the lumber industry as "copper-chromated-arsenate" (CCA), a fungicide and wood preservative. As of 2003 this product is no longer being used for wood preservation in the U.S., but still has some restricted uses. Due to its long history of use and persistence, it will be in our environment for years to come. With ingestion or high level dermal exposure, arsenic can produce a painful, predominantly sensory, peripheral polyneuropathy.

Other Agents

Vacor (N-3-pyridylmethy N′-p-nitrophenyl urea) is a rodenticide that produces a predominantly autonomic neuropathy.[67] Although it is banned in the U.S., it is mentioned due to its relatively selective effect on autonomic function.

CONCLUSION

Given the continuing and widespread use of cholinesterase inhibiting pesticides both in the U.S. and abroad, this group of compounds deserves special attention. There is evidence of persistent neurobehavioral and peripheral nervous system effects following high dose exposure, however the potential effects of low level, chronic exposure remain uncertain. Whether such exposures cause persistent neurological abnormalities is unclear, as attempts to answer this question have yielded inconsistent results. Studies have often suffered from the methodological issues, including problems with control matching or misclassification due to the lack of clear unexposed status among controls or during "pre-exposure" testing. Further research using refined exposure assessment tools and sensitive outcome measures will be needed in order to evaluate the effects of chronic, low level exposure, which is much more common than acute, high level exposure with clinically evident intoxication. Of special interest are workers potentially exposed to pesticide residue that are dislodged and may become bioavailable during food product harvesting processes.

The potential for abnormal neurodevelopmental effects of pesticide exposure are another particular concern, since the nervous system is known to be most susceptible to harm during developmentally sensitive periods. Certain pesticides may affect the fetus *in utero*,[68] indicating a potential for harm due to maternal exposures to pesticides.[69] Children may also be directly exposed to pesticides in their homes, as studies of house dust have revealed that pesticide residues can be found in the home environment.[70] These sources of exposure have special relevance for children, who spend more time on the floor and have more hand to mouth activity than adults.[71] Although concerted attention to an organic food diet and improved housekeeping practices appear to reduce pesticide and dust exposures,[72,73] the neurodevelopmental effects of these low-dose exposures remain uncertain. Whether exposure to these substances *in utero*, infancy, and childhood affect neurological development should be answered. Studies now underway may provide valuable information relating to these questions.

The nervous system represents a prime target for both the acute and chronic effects of pesticides. Organophosphate poisoning is a prototypical example, affecting both the central and peripheral nervous system with a spectrum of toxicity that includes an acute cholinergic crisis, an intermediate neuromuscular syndrome, and a chronic polyneuropathy. The potential effects of chronic, low-dose pesticide exposures on neurodevelopment remain uncertain. We have discussed only a few of the myriad of poisons which are directed at pests but have demonstrable toxicity in humans. Many new pesticides are specifically designed to poison the nervous system, so it is likely that their toxic effects on the human nervous system will become recognized over time.

REFERENCES

1. Costa LG. Basic toxicology of pesticides. Occup Med. 1997 Apr-Jun;12(2):251-68.

2. Reigart JR, Roberts JR. Recognition and Management of Pesticide Poisonings. Fifth Edition. Environmental Protection Agency, 1999 Washington DC. Available from: *http://www.epa.gov/pesticides/safety/healthcare/handbook/handbook.pdf* [cited 2006 Jun 15].

3. Cannon SB, Veazey JM Jr, Jackson RS, Burse VW, Hayes C, Straub WE, Landrigan PJ, Liddle JA. Epidemic kepone poisoning in chemical workers. Am J Epidemiol. 1978 Jun;107(6):529-37.

4. Taylor JR. Neurological manifestations in humans exposed to chlordecone and follow-up results. Neurotoxicology. 1982 Oct;3(2):9-16.

5. Anderson JM, Cockburn F, Forfar JO, Harkness RA, Kelly RW, Kilshaw B. Neonatal spongiform myelopathy after restricted application of hexachlorophane skin disinfectant. J Clin Pathol. 1981 Jan;34(1):25-9.

6. Slamovits TL, Burde RM, Klingele TG. Bilateral optic atrophy caused by chronic oral ingestion and topical application of hexachlorophene. Am J Opthamol. 1980 May; 89(5):676-9.

7. Narahashi T, Frey JM, Ginsburg KS, Roy ML. Sodium and GABA-activated channels as the targets of pyrethroids and cyclodienes. Toxicol Lett. 1992 Dec; 64-65 Spec No:429-36.

8. Savage EP, Keefe TJ, Mounce LM, Heaton RK, Lewis JA, Burcar PJ. Chronic neurological sequelae of acute organophosphate pesticide poisoning. Arch Environ Health. 1988 Jan-Feb;43(1):38-45.

9. Rosenstock L, Keifer M, Daniell WE, McConnell R, Claypoole K. Chronic central nervous system effects of acute organophosphate pesticide intoxication. The Pesticide Health Effects Study Group. Lancet. 1991 Jul 27;338(8761):223-7.

10. Steenland K, Jenkins B, Ames RG, O'Malley M, Chrislip D, Russo J. Chronic neurological sequelae to organophosphate pesticide poisoning. Am J Public Health. 1994 May;84(5):731-6.

11. Wesseling C, Keifer M, Ahlbom A, McConnell R, Moon JD, Rosenstock L, Hogstedt C. Long-term neurobehavioral effects of mild poisonings with organophosphate and n-methyl carbamate pesticides among banana workers. Int J Occup Environ Health. 2002 Jan-Mar;8(1):27-34.

12. Rosenstock L, Daniell W, Barnhart S, Schwartz D, Demers PA. Chronic neuropsychological sequelae of occupational exposure to organophosphate insecticides. Am J Ind Med. 1990;18(3):321-5.

13. Stallones L, Beseler C. Pesticide illness, farm practices, and neurological symptoms among farm residents in Colorado. Environ Res. 2002 Oct;90(2):89-97.

14. Stephens R, Spurgeon A, Calvert IA, Beach J, Levy LS, Berry H, Harrington JM. Neuropsychological effects of long-term exposure to organophosphates in sheep dip. Lancet. 1995 May 6;345(8958):1135-9.

15. Roldan-Tapia L, Parron T, Sanchez-Santed F. Neuropsychological effects of long-term exposure to organophosphate pesticides. Neurotoxicol Teratol. 2005 Mar-Apr;27(2):259-66.

16. Steenland K, Dick RB, Howell RJ, Chrislip DW, Hines CJ, Reid TM, Lehman E, Laber P, Krieg EF Jr, Knott C. Neurologic function among termiticide applicators exposed to chlorpyrifos. Environ Health Perspect. 2000 Apr;108(4):293-300.

17. Fiedler N, Kipen H, Kelly-McNeil K, Fenske R. Long-term use of organophosphates and neuropsychological performance. Am J Ind Med. 1997 Nov;32(5): 487-96.

18. Ames RG, Steenland K, Jenkins B, Chrislip D, Russo J. Chronic neurologic sequelae to cholinesterase inhibition among agricultural pesticide applicators. Arch Environ Health. 1995 Nov-Dec;50(6):440-4.

19. Burgess JL, Morrissey B, Keifer MC, Robertson WO. Fumigant-related illnesses: Washington State's five-year experience. J Toxicol Clin Toxicol. 2000;38(1): 7-14.

20. Maddy KT, Edmiston S, Richmond D. Illness, injuries, and deaths from pesticide exposures in Califor-

nia 1949-1988. Rev Environ Contam Toxicol. 1990; 114:57-123.

21. Kishi R, Itoh I, Ishizu S, Harabuchi I, Miyake H. Symptoms among workers with long-term exposure to methyl bromide. An epidemiological study. Sangyo Igaku. 1991 Jul;33(4):241-50.

22. Bishop CM. A case of methyl bromide poisoning. Occup Med (Lond). 1992 May;42(2):107-9.

23. De Haro L, Gastaut JL, Jouglard J, Renacco E. Central and peripheral neurotoxic effects of chronic methyl bromide intoxication. J Toxicol Clin Toxicol. 1997;35(1):29-34.

24. Reidy TJ, Bolter JF, Cone JE. Neuropsychological sequelae of methyl bromide: a case study. Brain Inj. 1994 Jan;8(1):83-93.

25. Brashear A, Unverzagt FW, Farber MO, Bonnin JM, Garcia JG, Grober E. Ethylene oxide neurotoxicity: a cluster of 12 nurses with peripheral and central nervous system toxicity. Neurology. 1996 Apr;46(4):992-8.

26. Estrin WJ, Bowler RM, Lash A, Becker CE. Neurotoxicological evaluation of hospital sterilizer workers exposed to ethylene oxide. J Toxicol Clin Toxicol. 1990;28(1):1-20.

27. Centers for Disease Control (CDC). Fatalities resulting from sulfuryl fluoride exposure after home fumigation–Virginia. MMWR Morb Mortal Wkly Rep. 1987 Sep 18;36(36):602-4, 609-11.

28. Scheuerman EH. Suicide by exposure to sulfuryl fluoride. J Forensic Sci. 1986 Jul;31(3):1154-8.

29. Calvert GM, Mueller CA, Fajen JM, Chrislip DW, Russo J, Briggle T, Fleming LE, Suruda AJ, Steenland K. Health effects associated with sulfuryl fluoride and methyl bromide exposure among structural fumigation workers. Am J Public Health. 1998 Dec;88(12):1774-80.

30. Anger WK, Moody L, Burg J, Brightwell WS, Taylor BJ, Russo JM, Dickerson N, Setzer JV, Johnson BL, Hicks K. Neurobehavioral evaluation of soil and structural fumigators using methyl bromide and sulfuryl fluoride. Neurotoxicology. 1986 Fall;7(3):137-56.

31. Agin H, Calkavur S, Uzun H, Bak M. Amitraz poisoning: clinical and laboratory findings. Indian Pediatr. 2004 May;41(5):482-6.

32. Yilmaz HL, Yildizdas DR. Amitraz poisoning, an emerging problem: epidemiology, clinical features, management, and preventive strategies. Arch Dis Child. 2003 Feb;88(2):130-4.

33. Kalyoncu M, Dilber E, Okten A. Amitraz intoxication in children in the rural Black Sea region: analysis of forty-three patients. Hum Exp Toxicol. 2002 May; 21(5):269-72.

34. National Pesticide Telecommunications Network web site. Fipronil. Available from: *http://npic.orst.edu/ factsheets/fipronil.pdf* [cited 2006 Jun 19].

35. Mohamed F, Senarathna L, Percy A, Abeyewardene M, Eaglesham G, Cheng R, Azher S, Hittarage A, Dissanayake W, Sheriff MH, Davies W, Buckley NA, Eddleston M. Acute human self-poisoning with the N-phenylpyrazole insecticide fipronil-a GABAA-gated chloride channel blocker. J Toxicol Clin Toxicol. 2004; 42(7):955-63.

36. Priyadarshi A, Khuder SA, Schaub EA, Shrivastava S. A meta-analysis of Parkinson's disease and exposure to pesticides. Neurotoxicology. 2000 Aug; 21(4):435-40.

37. Yang WL, Sun AY. Paraquat-induced cell death in PC12 cells. Neurochem Res. 1998 Nov;23(11):1387-94.

38. Uversky VN, Li J, Bower K, Fink AL. Synergistic effects of pesticides and metals on the fibrillation of alpha-synuclein: implications for Parkinson's disease. Neurotoxicology. 2002 Oct;23(4-5):527-36.

39. Betarbet R, Sherer TB, MacKenzie G, Garcia-Osuna M, Panov AV, Greenamyre JT. Chronic systemic pesticide exposure reproduces features of Parkinson's disease. Nat Neurosci. 2000 Dec;3(12):1301-6.

40. McCormack AL, Thiruchelvam M, Manning-Bog AB, Thiffault C, Langston JW, Cory-Slechta DA, Di Monte DA. Environmental risk factors and Parkinson's disease: selective degeneration of nigral dopaminergic neurons caused by the herbicide paraquat. Neurobiol Dis. 2002 Jul;10(2):119-27.

41. Bhatt MH, Elias MA, Mankodi AK. Acute and reversible parkinsonism due to organophosphate pesticide intoxication: five cases. Neurology. 1999 Apr 22; 52(7):1467-71.

42. Meco G, Bonifati V, Vanacore N, Fabrizio E. Parkinsonism after chronic exposure to the fungicide maneb (manganese ethylene-bis-dithiocarbamate). Scand J Work Environ Health. 1994 Aug;20(4):301-5.

43. Rajput AH, Uitti RJ, Stern W, Laverty W. Early onset Parkinson's disease in Saskatchewan–environmental considerations for etiology. Can J Neurol Sci. 1986 Nov;13(4):312-6.

44. Barbeau A, Roy M, Bernier G, Campanella G, Paris S. Ecogenetics of Parkinson's disease: prevalence and environmental aspects in rural areas. Can J Neurol Sci. 1987 Feb;14(1):36-41.

45. Koller W, Vetere-Overfield B, Gray C, Alexander C, Chin T, Dolezal J, Hassanein R, Tanner C. Environmental risk factors in Parkinson's disease. Neurology. 1990 Aug;40(8):1218-21.

46. Liou HH, Tsai MC, Chen CJ, Jeng JS, Chang YC, Chen SY, Chen RC. Environmental risk factors and Parkinson's disease: a case-control study in Taiwan. Neurology. 1997 Jun;48(6):1583-8.

47. Firestone JA, Smith-Weller T, Franklin G, Swanson P, Longstreth WT Jr, Checkoway H. Pesticides and risk of Parkinson disease: a population-based case-control study. Arch Neurol. 2005 Jan;62(1):91-5.

48. Gorell JM, Johnson CC, Rybicki BA, Peterson EL, Richardson RJ. The risk of Parkinson's disease with exposure to pesticides, farming, well water, and rural living. Neurology. 1998 May;50(5):1346-50.

49. Semchuk KM, Love EJ, Lee RG. Parkinson's disease and exposure to agricultural work and pesticide chemicals. Neurology. 1992 Jul; 42(7):1328-35.

50. Hertzman C, Wiens M, Snow B, Kelly S, Calne D. A case-control study of Parkinson's disease in a hor-

ticultural region of British Columbia. Mov Disord. 1994 Jan;9(1):69-75.

51. Baldi I, Lebailly P, Mohammed-Brahim B, Letenneur L, Dartigues JF, Brochard P. Neurodegenerative diseases and exposure to pesticides in the elderly. Am J Epidemiol. 2003 Mar;157(5):409-14.

52. Kamel F, Engel LS, Gladen BC, Hoppin JA, Alavanja MC, Sandler DP. Neurologic symptoms in licensed private pesticide applicators in the agricultural health study. Environ Health Perspect. 2005 Jul;113(7): 877-82.

53. Miranda J, Lundberg I, McConnell R, Delgado E, Cuadra R, Torres E, Wesseling C, Keifer M. Onset of grip- and pinch-strength impairment after acute poisonings with organophosphate insecticides. Int J Occup Environ Health. 2002 Jan-Mar;8(1):19-26.

54. Miranda J, McConnell R, Wesseling C, Cuadra R, Delgado E, Torres E, Keifer M, Lundberg I. Muscular strength and vibration thresholds during two years after acute poisoning with organophosphate insecticides. Occup Environ Med. 2004 Jan;61(1):e4.

55. Lotti M, Moretto A. Organophosphate-induced delayed polyneuropathy. Toxicol Rev. 2005;24(1):37-49.

56. Johnson MK. Sensitivity and selectivity of compounds interacting with neuropathy target esterase. Further structure-activity studies. Biochem Pharmacol. 1988 Nov 1;37(21):4095-104.

57. De Bleecker J, Van Den Neucker K, Willems J. The intermediate syndrome in organophosphate poisoning: presentation of a case and review of the literature. J Toxicol Clin Toxicol. 1992;30(3):321-9; discussion 331-2.

58. Sudakin DL, Mullins ME, Horowitz BZ, Abshier V, Letzig L. Intermediate syndrome after malathion ingestion despite continuous infusion of pralidoxime. J Toxicol Clin Toxicol. 2000;38(1):47-50.

59. Tomizawa M, Casida JE. Selective toxicity of neonicotinoids attributable to specificity of insect and mammalian nicotinic receptors. Annu Rev Entomol. 2003;48:339-64.

60. Wu IW, Lin JL, Cheng ET. Acute poisoning with the neonicotinoid insecticide imidacloprid in N-methyl pyrrolidone. J Toxicol Clin Toxicol. 2001;39(6):617-21.

61. Bradberry SM, Watt BE, Proudfoot AT, Vale JA. Mechanisms of toxicity, clinical features, and management of acute chlorophenoxy herbicide poisoning: a review. J Toxicol Clin Toxicol. 2000;38(2):111-22.

62. Lifshitz M, Gavrilov V. Central nervous system toxicity and early peripheral neuropathy following dermal exposure to methyl bromide. J Toxicol Clin Toxicol. 2000;38(7):799-801.

63. Yang RS, Witt KL, Alden CJ, Cockerham LG. Toxicology of methyl bromide. Rev Environ Contam Toxicol. 1995;142:65-85.

64. Kuzuhara S, Kanazawa I, Nakanishi T, Egashira T. Ethylene oxide polyneuropathy. Neurology. 1983 Mar;33(3):377-80.

65. Saddique A, Peterson CD. Thallium poisoning: a review. Vet Hum Toxicol. 1983 Feb;25(1):16-22.

66. Desenclos JC, Wilder MH, Coppenger GW, Sherin K, Tiller R, VanHook RM. Thallium poisoning: an outbreak in Florida, 1988. South Med J. 1992 Dec;85(12): 1203-6.

67. LeWitt PA. The neurotoxicity of the rat poison vacor. A clinical study of 12 cases. N Engl J Med. 1980 Jan 10;302(2):73-7.

68. Souza MS, Magnarelli GG, Rovedatti MG, Cruz SS, De D'Angelo AM. Prenatal exposure to pesticides: analysis of human placental acetylcholinesterase, glutathione S-transferase and catalase as biomarkers of effect. Biomarkers. 2005;10(5):376-89.

69. Young JG, Eskenazi B, Gladstone EA, Bradman A, Pedersen L, Johnson C, Barr DB, Furlong CE, Holland NT. Association between in utero organophosphate pesticide exposure and abnormal reflexes in neonates. Neurotoxicology. 2005;26(2):199-209.

70. Fenske RA, Lu C, Barr D, Needham L. Children's exposure to chlorpyrifos and parathion in an agricultural community in central Washington State. Environ Health Perspect. 2002;110(5):549-53.

71. Curl CL, Fenske RA, Kissel JC, Shirai JH, Moate TF, Griffith W, Coronado G, Thompson B. Evaluation of take-home organophosphorus pesticide exposure among agricultural workers and their children. Environ Health Perspect. 2002;110(12):A787-92.

72. Curl CL, Fenske RA, Elgethun K. Organophosphorus pesticide exposure of urban and suburban preschool children with organic and conventional diets. Environ Health Perspect. 2003;111(3):377-82.

73. Lu C, Toepel K, Irish R, Fenske RA, Barr DB, Bravo R. Organic diets significantly lower children's dietary exposure to organophosphorus pesticides. Environ Health Perspect. 2006;114(2):260-3.

doi:10.1300/J096v12n01_03

Reproductive Disorders Associated with Pesticide Exposure

Linda M. Frazier, MD, MPH

SUMMARY. Exposure of men or women to certain pesticides at sufficient doses may increase the risk for sperm abnormalities, decreased fertility, a deficit of male children, spontaneous abortion, birth defects or fetal growth retardation. Pesticides from workplace or environmental exposures enter breast milk. Certain pesticides have been linked to developmental neurobehavioral problems, altered function of immune cells and possibly childhood leukemia. In well-designed epidemiologic studies, adverse reproductive or developmental effects have been associated with mixed pesticide exposure in occupational settings, particularly when personal protective equipment is not used. Every class of pesticides has at least one agent capable of affecting a reproductive or developmental endpoint in laboratory animals or people, including organophosphates, carbamates, pyrethroids, herbicides, fungicides, fumigants and especially organochlorines. Many of the most toxic pesticides have been banned or restricted in developed nations, but high exposures to these agents are still occurring in the most impoverished countries around the globe. Protective clothing, masks and gloves are more difficult to tolerate in hot, humid weather, or may be unavailable or unaffordable. Counseling patients who are concerned about reproductive and developmental effects of pesticides often involves helping them assess their exposure levels, weigh risks and benefits, and adopt practices to reduce or eliminate their absorbed dose. Patients may not realize that by the first prenatal care visit, most disruptions of organogenesis have already occurred. Planning ahead provides the best chance of lowering risk from pesticides and remediating other risk factors before conception. doi:10.1300/J096v12n01_04 *[Article copies available for a fee from The Haworth Document Delivery Service: 1-800-HAWORTH. E-mail address: <docdelivery@haworthpress.com> Website: <http://www.HaworthPress.com> © 2007 by The Haworth Press, Inc. All rights reserved.]*

KEYWORDS. Pesticides, reproduction, pregnancy, human development, fertility, abortion, spontaneous, teratogens, fetal growth retardation, breast feeding, neurotoxicity syndromes

During the twentieth century, more than 50,000 pesticide formulations were used to combat insects, nematodes, fungi and unwanted plants. These products helped to increase crop yields, lower food costs and reduce deaths from vector-borne diseases. In recent decades, significant human toxicity from certain pesticides was documented, and agro-

Linda M. Frazier is affiliated with the Departments of Obstetrics and Gynecology, University of Kansas School of Medicine, Wichita, KS.

Address correspondence to: Linda M. Frazier, MD, MPH, Departments of Obstetrics and Gynecology, University of Kansas School of Medicine-Wichita, 1010 North Kansas Avenue, Wichita, KS 67214.

[Haworth co-indexing entry note]: "Reproductive Disorders Associated with Pesticide Exposure." Frazier, Linda M. Co-published simultaneously in *Journal of Agromedicine* (The Haworth Medical Press, an imprint of The Haworth Press, Inc.) Vol. 12, No. 1, 2007, pp. 27-37; and: *Proceedings from the Medical Workshop on Pesticide-Related Illnesses from the International Conference on Pesticide Exposure and Health* (ed: Ana Maria Osorio, and Lynn R. Goldman) The Haworth Medical Press, an imprint of The Haworth Press, Inc., 2007, pp. 27-37. Single or multiple copies of this article are available for a fee from The Haworth Document Delivery Service [1-800-HAWORTH, 9:00 a.m. - 5:00 p.m. (EST). E-mail address: docdelivery@haworthpress.com].

Available online at http://ja.haworthpress.com
© 2007 by The Haworth Press, Inc. All rights reserved.
doi:10.1300/J096v12n01_04

chemical research led to the development of less toxic alternatives. People have environmental exposures to pesticides through dietary intake and other routes, resulting in measurable but usually low levels in the body. This is particularly true for organochlorine compounds used as pesticides because they have long half-lives in humans, animals and the environment.

Symptoms from poisoning after applying pesticides that have anticholinergic effects (e.g., organophosphates and carbamates) appear relatively rapidly. The large demand for educational materials that summarize the acute toxicity of pesticides is illustrated by the publication of five editions of the Environmental Protection Agency's *Recognition and Management of Pesticide Poisonings,*[1] including an edition in Spanish. Health problems from pesticides in the absence of acute poisoning are also clinically important. This article reviews examples of reproductive disorders that may be diagnosed weeks, months or years later after exposure to agents from each of the major classes of pesticides.

EXPOSURES AMONG MEN

Reproductive hazards in the environment or workplace may affect either men or women (Table 1). Problems resulting from a man's pesticide exposures may be caused by alterations in the structure or function of sperm genetic material,[2-15] or by other toxic damage to sperm or testicular function.[16-21] A man who has circulating blood levels of pesticides may theoretically expose a woman's uterine environment during intercourse by depositing pesticide-laden semen in the vagina. It has been well documented that family members of pesticide-exposed men can be exposed when the home is contaminated with these agrochemicals from his clothing, skin or shoes.[22] Epidemiologic studies of agricultural workers and studies among laboratory animals suggest that male exposures are associated with a variety of adverse reproductive effects (Table 2).

Reduced Fertility: Infertility among men exposed to the fumigant dibromochloropropane (DBCP) provided a dramatic example in the 1970s of men's vulnerability to testicular toxicants.[20,21] Nearly half of 107 male manufacturing workers who handled DBCP in a California facility had low sperm concentrations, and most of the affected men had azoospermia or severe oligospermia. After following these men and a similar cohort from Israel for many years after cessation of exposure, poor sperm indices commonly persisted. No other pesticides have been found to have such striking effects on male fertility, although in one study, reduced fecundity was noted among men who were exposed to pyrethroids, organophosphates and carbamates in greenhouses.[23] A causal relationship was supported by the fact that greenhouse workers who did not use personal protective equipment were the only workers who experienced fertility problems. Male rats exposed *in utero* during gonadal sex determination to the estrogenic organochlorine insecticide, methoxychlor, or the antiandrogenic fungicide, vinclozolin, had decreased sperm concentrations.[11] Reduced sperm concentrations present in their sons suggested that endocrine disruptors could act transgenerationally by an epigenetic mechanism.

Deficit of Male Children: There were fewer male than female infants among DBCP-exposed men who were later able to conceive.[20,21] An abnormal sex ratio with fewer male children has also been noted among fathers who were pesticide applicators, or greenhouse workers who predominantly used fungicides.[23-25]

Spontaneous Abortion: Early fetal death can be related to a man's pesticide use according to findings in epidemiologic studies of pesticide

TABLE 1. Selected Reproductive Health Problems After Exposure to Pesticides Among Men or Women

Problem	Partner Exposed
Reduced Fertility	Woman, Man
Abnormal Sperm Genetic Material or Indices	Man
Deficit of Male Children	Man
Spontaneous Abortion	Woman, Man
Birth Defects	Woman, Man
Fetal Growth Retardation, Preterm Birth	Woman
Breast Milk Contamination	Woman
Childhood Neurobehavioral Problems	Woman
Developmental Immunotoxicity	Woman
Childhood Cancer	Possibly Woman

TABLE 2. Male Exposures Associated with Adverse Reproductive Effects

Pesticide Class	Example Agents	Study Details	References
Mixed Exposure	Wide variety	Sperm aneuploidy in agricultural workers, decreased fertility and fewer male offspring in men working in greenhouses. Spontaneous abortions in wives and birth defects in children of farmers or pesticide applicators.	2, 23-29
Organophosphates	Ethyl parathion, methamidophos, chlorpyrifos, diazinon	Sperm aneuploidy in manufacturing workers and sperm DNA fragmentation in applicators (parathion, methamdidophos); sperm DNA fragmentation in men treated for infertility (chlorpyrifos). DNA fragmentation and altered sperm viability, motility, morhphology in mice (diazinon).	3-6
Carbamates	Carbaryl, carbosulfan, carbofuran	Sperm aneuploidy and abnormal morphology in manufacturing workers (carbaryl). DNA fragmentation in men treated for infertility (carbaryl). Chromosome aberrations and abnormal head morphology in sperm of mice (carbosulfan, carbofuran).	5, 7, 8, 16
Pyrethroids	Fenvalerate, cypermethrin	Sperm aneuploidy and DNA fragmentation in manufacturing workers (fenvalerate). Sperm indices and reproductive hormone levels altered in rats and rabbits (cypermethrin).	9, 10, 17, 18
Herbicides	Paraquat, chlorophenoxy herbicides, triazines, glyphosate	Sperm shape abnormalities in mice (paraquat). Spontaneous abortions in farm families using certain herbicides.	19, 26
Fungicides	Vinclozolin, other fungicides	Decreased sperm concentration after in-utero exposure in rats (vinclozolin). Fewer male children among fungicide applicators and greenhouse workers.	11, 23, 25
Fumigants	DBCP,* methyl bromide	Irreversible azoospermia, fewer male children in manufacturing workers and applicators, and DNA strand breaks in male rat germ cells (DBCP*). Germ cell chromosome mutations (methyl bromide).	12, 13, 20, 21
Organochlorines	DDT* metabolites, lindane, methoxychlor	Slightly increased sperm DNA fragmentation in fishermen (DDT*). Sperm chromatin abnormalities in mice (lindane) and decreased sperm concentration in rats (methoxychlor) after *in utero* exposure.	11, 14, 15

* DBCP, dibromochloropropane. DDT, dichlorodiphenyltrichloroethane.

applicators in Minnesota and farm families in Ontario, Canada.[24,26,27] The studies were large (695 and 2,110 families, respectively) and controlled for multiple potential confounders. Particularly vulnerable were couples already at risk for spontaneous abortion because the woman was older than 34 years.[26] Although the studies were retrospective, two factors supported causality. Most miscarriages occurred in spring (when herbicides were used), and risk was highest if the husband did not wear protective equipment when applying pesticides.

The known mechanisms that lead to spontaneous abortion support the biologic plausibility of a causal relationship with paternal pesticide exposure. In the general population, about one-third of spontaneous abortions are aneuploid, frequently involving chromosome18.[2] The extra chromosome implicated in miscarriages can be of paternal origin. Several prospective studies examined aneuploidy in the sperm of pesticide applicators or pesticide manufacturing workers.[2,3,7,9] Occupational exposure to insecticides were confirmed by urinary metabolite assays, personal air samples or dermal expo-

sure measurements. Abnormal chromosome numbers were found in the sperm of these men. There was an increase in sex null aneuploidy associated with organophosphates,[2] and an elevated frequency of disomy in chromosomes X, Y and 18. The exposures were to ethyl parathion, methamidophos, carbaryl and the pyrethroid, fenvalerate.[3,7,9]

Birth Defects: Birth defects have been linked to paternal exposure to pesticides in several large epidemiologic studies.[25,28,29] This possible risk is supported somewhat by studies showing disruption of genetic processes in the sperm of humans and animals on exposure to many types of pesticides (Table 2). Although the epidemiologic studies were potentially limited by recall bias, balancing this limitation were use of experts to assign likely exposure levels based on detailed questionnaires, and analyses that controlled for confounding and considered Mendelian inheritance patterns. Even though no single type of anomaly was predominant, the two studies independently found associations of birth defects with herbicide use. Suggesting that further research may be warranted to explore this potential cause of birth defects was the timing of paternal exposure, in that children of herbicide users were more likely to have birth defects if conceived in the spring.[25]

EXPOSURES AMONG WOMEN

When women are sufficiently exposed to certain pesticides, several types of adverse reproductive outcomes may occur, and developmental problems in their children may result (Table 1). In addition to the relatively low levels of environmental exposure from sources such as fish, fruit and vegetables, women can be occupationally exposed to large doses of pesti-

cides as applicators, farm workers and in other jobs. In the Agricultural Health Study, 1,359 female licensed pesticide applicators were enrolled from just two states (North Carolina and Iowa).[30] In developing countries, it is common for women to perform chemically-intensive agricultural tasks.[31] Examples of adverse effects that have been associated with various types of pesticides are summarized on Table 3.

Reduced Fertility: Take-home toxicants may have played a role in the study where wives of male greenhouse workers required a longer time period to become pregnant.[25] In a study that compared 322 women presenting for infertility care (cases) with an equal number of controls, mixing and applying herbicides and use of fungicides were much more common among

TABLE 3. Female Exposures Associated with Adverse Reproductive Effects

Pesticide Class	Example Agents	Study Details	References
Mixed Exposure	Wide variety	Longer time to pregnancy among women whose spouses work in greenhouses. Spontaneous abortions in wives and birth defects in children of farmers or pesticide applicators. Women with infertility more likely than controls to be exposed to pesticides. Risk of childhood leukemia after indoor pesticide exposure during pregnancy. Formulations may include organic solvents that exhibit reproductive or developmental toxicity.	23-29, 33, 58
Organophosphates	Chlorpyrifos, diazinon, malathion, parathion	Lower birth weight and shorter gestational age among women with residential or agricultural exposure (agents listed at left). Neurodevelopmental or childhood behavioral problems (chlorpyrifos, diazinon). Altered fetal immune cell function in rats (chlorpyrifos).	44, 45, 56
Carbamates	Carbaryl, propoxur	Neurodevelopmental or childhood behavioral problems (carbaryl). Possibly childhood leukemia (propoxur).	51, 59
Pyrethroids	Deltamethrin, permethrin	Neurodevelopmental or childhood behavioral problems (deltamethrin). Possibly childhood leukemia (very high exposure to permethrin).	51, 60
Herbicides	Chlorophenoxy herbicides, triazines, glyphosate	Women with infertility more likely to mix and apply herbicides. Spontaneous abortions in farm families using certain herbicides. Limb and heart defects in children of pesticide applicators or agricultural workers using herbicides. Early pregnancy loss and altered fetal immune cell function in rats, possibly human fetal growth restriction from community drinking water (atrazine).	25, 26, 33, 37, 39-42, 46, 56
Fungicides	Vinclozolin, thiram	Women with infertility more likely to use fungicides, supported by some studies in mice and rats. Limb defects in children of pesticide applicators using fungicides. Pregnant rats exposed during gonadal sex determination had male offspring with decreased sperm concentration that also affected subsequent generations of males (vinclozolin).	11, 33, 35, 40
Fumigants		Data inadequate (see text related to neurodevelopment).	
Organochlorines	Dioxin, methoxychlor, DDT,* others	Increased rates of endometriosis among women in Seveso, Italy (dioxin), supported by studies in mice, rats and monkeys. Fetal growth restriction in women with diets high in contaminated fish, or who have higher blood levels of chlorinated pesticides (DDT* and others). Pregnant rats exposed during gonadal sex determination had male offspring with decreased sperm concentration that also affected subsequent generations of males (methoxychlor). Decreased immune cell function at birth and increased middle ear infections after human *in utero* exposure.	11, 36, 47, 54, 55

* DDT, dichlorodiphenyltrichloroethane.

the case women.[32] Other epidemiologic studies suggest that human fertility may be reduced by pesticide use, and disrupted development of preimplantation embryos in mice by low doses of commonly-used herbicides, insecticides and fungicides has been reported.[33] In whole animal systems, however, one study showed no impact on fertility from mixtures of agrochemicals at doses 100-fold greater than those found in contaminated groundwater.[34] In another study, the fungicide, thiram, reduced fertilization of oocytes in rats.[35] Among the oocytes that were fertilized, many were polyspermic, leading to polyploidy, a common cause of miscarriage.

Reduced fertility can be the result of endometriosis. This disorder has been linked to dioxin exposure in women exposed in Seveso, Italy, although the increase was not statistically significant.[36] Dioxin has a toxicologic profile that is similar to some persistent organochlorine pesticides. Endometriosis is caused by dioxin in mice, rats and monkeys, particularly if exposure begins *in utero*.

Spontaneous Abortion: Unrecognized spontaneous abortions early in gestation can manifest as reduced fertility. Clinically-apparent spontaneous abortions were increased among farm families using phenoxy acid herbicides, glyphosate or triazines in one study.[26] Lending support to this study is the finding that the triazine herbicide, atrazine, induces early pregnancy loss in rats due to endocrine disruption.[37] No increase in spontaneous abortion was found after malathion was extensively sprayed to combat fruit flies in California.[38] Some case-control studies have suggested that low level DDT exposure may lead to miscarriages, but the research is not conclusive.[39]

Birth Defects: Use of pesticides that induce major birth defects in the absence of maternal toxicity in rodents is typically restricted in developed countries. Despite this, congenital anomalies appear to be increased by certain pesticides in epidemiologic studies. This suggests that either the epidemiologic findings are spurious due to confounding or bias, or that standard rodent test protocols may not always predict human responses. Well-designed studies by several epidemiology teams in different countries have found associations between pesticide use and limb reduction anomalies, and sometimes congenital heart defects and other anomalies.[40-43] One of the largest investigations linked records of 210,723 births in Minnesota with the list of state-licensed pesticide applicators.[40] In regions where chlorophenoxy herbicides and fungicides were used extensively on crops, there were 30.0 infants with anomalies per thousand live births among applicator families, compared to 26.9 and 18.3 per thousand among the general public in Minnesota's crop and non-crop regions, respectively. Birth defect rates in the crop regions were most elevated for infants conceived in the spring, and there was no seasonal increase in the anomalies in non-crop regions.

Fetal Growth Retardation or Preterm Birth: Infant birth weight averaged 215 grams less among pregnant women who had high-level exposures to the insecticides chlorpyrifos and diazinon from treatment of their home for roaches and other insects.[44] This reduction in birth weight was of the same magnitude as the fetal growth restriction that occurs as a result of smoking. Insecticide levels were measured in maternal and newborn blood at delivery. A similar study found a slight increase in preterm birth related to chlorpyrifos and other organophosphates.[45] Fetal growth restriction was associated with modest exposures from atrazine-contaminated drinking water in another community, although the study was not conclusive.[46] Several epidemiologic studies in Sweden have shown that women with a high dietary intake of fish contaminated with organochlorines have infants with lower birth weights, and a study from India confirmed this association using blood and placenta assays for DDT and similar organochlorine compounds.[47]

Breast Milk Contamination: Pesticides enter into breast milk not only from acute maternal exposures, but also from a woman's body stores.[48,49] Pesticides with the highest body burden are typically from the organochlorine family and include lindane, aldrin, dieldrin, endosulfan, mirex, chlordane, and DDT. Breast fed infants clearly have higher organochlorine levels in their blood than bottle-fed infants,[50] although experts are not sure whether risks from these levels outweigh the benefits of breast feeding. One team has estimated that if a mother carries a high body burden of organochlorines, her infant may accumulate during six months of breast feeding as much exposure as an adult would have in 25 years.[48]

DDT levels in breast milk have fallen in countries that banned its use–the United States enacted such a ban in 1972. This insecticide is still being used in developing countries, and the following high breast milk levels have been reported (as concentration in milk fat): Hong Kong 13,800 µg/kg (1985), Kenya 4,800 µg/kg (1986), South Africa 15,830 µg/kg (1987), Vietnam 11,400 µg/kg (1989), Zimbabwe 6,000 µg/kg (1990), and Mexico 6,440 µg/kg (1995).[49] Nursing infants exceed the World Health Organization's allowable daily intake for adults at breast milk DDT concentrations of 5,000-6,000 µg/kg in milk fat. The intake for infants may need to be even lower than this because babies may be more sensitive to toxic effects than adults.

Childhood Neurobehavioral Problems: Exposure to pesticides such as carbaryl, deltamethrin, diazinon and chlorpyrifos during fetal development has been linked to neurodevelopmental disorders, sometimes called behavioral teratogenicity, in several animal species. Chlorpyrifos impairs development of cellular structures in the brain, learning and behavior in rats at doses that do not cause adverse effects in the pregnant females.[51] These neurobehavioral abnormalities are evident when exposure occurs as early as the period of neural tube formation, which begins about three weeks after conception in humans (Table 4). Rodents exposed to chlorpyrifos *in utero* or shortly after birth may appear to recover from neurotoxicity in the short term but if followed through adolescence, deficits in brain cell numbers, synaptic communication and performance on behavioral tests are seen.[51]

TABLE 4. Critical Periods During Early Pregnancy for Development of Major Structural Birth Defects.

Organ	Weeks of Gestation
Heart	3 to 8
Central Nervous System	3 to 16
Limbs	4 to 8
Kidneys	4 to 16
Palate	6 to 10

Historically, learning, memory and behavioral tests were not included routinely in reproductive toxicology protocols that were required when many pesticides were evaluated by the Environmental Protection Agency. Based in part on research showing neurodevelopmental effects of chlorpyrifos, use of this pesticide in many home and agricultural settings in the United States has now been discontinued.

Other pesticides that are neurotoxic in adults have not been as extensively tested for effects on learning, memory or behavior. The fumigant methyl bromide, for example, could theoretically cause neurobehavioral problems after substantial *in utero* exposure. High exposure during fetal development of brain structures to pesticide formulations using neurotoxic organic solvents could also theoretically increase risk for neurobehavioral problems.

Mosquito repellents containing *N, N*-diethyl-*m*-toluamide (DEET) are recommended to prevent serious infections from West Nile virus and malaria. When repellents containing a high percentage by weight of DEET are applied to a large area of skin among infants and toddlers, neurologic symptoms can result. In a study of 449 pregnant women in Thailand who applied DEET regularly, DEET crossed the placenta and was detected in 8% of cord blood samples.[52] Compared to 448 women who used a repellent without DEET, there was no increase in neurologic abnormalities detected by a basic clinical examination when the infants were evaluated at birth and one year of age.

Developmental Immunotoxicity: The increased prevalence of atopic illness in developed countries is not well understood, but a reduction in exposure to infectious agents is thought to be a more likely explanation than toxic effects of chemicals on the immune system.[53] Prenatal exposure to DDT metabolites and organochlorine compounds such as dioxins and PCBs has been linked to decreased immune cell function in human cord blood and increased middle ear infections in children.[54,55] Studies in rodents suggest that *in utero* exposure to chlorpyrifos or atrazine may also alter results of tests on immune system cells.[56] Use of permethrin-treated bed nets in a malaria-endemic region in Kenya was associated with a reduction in two malaria antibodies in both maternal and cord blood.[57] Other antibody levels were preserved, however, and bed nets lowered morbidity by decreasing the rate of clinical malaria.

Childhood Cancer: Exposure to indoor pesticides during pregnancy doubled the risk for childhood leukemia in a well-designed case-control study in California.[58] At the time of the study, the most common home insecticides used in the region were chlorpyrifos, piperonyl butoxide, pyrethrins and propoxur. In a study of 136 cases of infant leukemia and 266 controls, having an affected child was associated with maternal exposure to pesticides during pregnancy, particularly the carbamate, propoxur.[59] The finding was most striking among infants with cleavage of the mixed lineage leukemia (*MLL*) gene, which is a known mechanism of leukemia induction after treatment with antineoplastic drugs or radiation. A case of infant leukemia with *MLL* gene rearrangement has also been reported after intensive exposure to permethrin during pregnancy.[60]

PATIENT MANAGEMENT

Counseling patients who are concerned about reproductive and developmental effects of pesticides often involves helping them weigh risks and benefits. There are situations when the benefits outweigh the risks of using a pesticide that can slightly increase the chance of a reproductive or developmental problem. A good example is using permethrin-treated bed nets to prevent malaria.[57] Similarly, applying a repellent containing DEET during pregnancy may be warranted when a woman cannot avoid contact with mosquitoes carrying a serious disease. The Centers for Disease Control and Prevention recommends that pregnant women at risk for contracting a major mosquito-borne disease should limit time in the out of doors and use protective clothing. Pregnant women, as with other adults, should apply an insect repellent containing DEET, but use it sparingly. During pregnancy, DEET should only be used when outdoors, and the repellent should be washed off with soap and water after coming indoors.

It is clear that the timing of exposure is very important. Most studies suggest that exposures in the few months before conception convey the most risk for fertility problems. The periconceptual time interval is also a key period of risk for miscarriages and birth defects. Planning ahead provides the best chance of lowering risk for birth defects by reducing exposures before conception and through the first trimester. Many patients do not realize that by the first prenatal care visit, most fetal disturbances of organogenesis have already occurred (Table 4).[61] Before becoming pregnant, other risk factors should also be reduced, including tobacco, teratogenic medications, low folate intake and others.[62, 63]

Even when there are no known hazardous exposures, patients may experience a reproductive disorder. Fertility problems occur in about 1 in 6 couples, and male factor abnormalities contribute to the problem in 30 to 50% of infertile couples seeking treatment. An occupational and environmental exposure history is part of preconception counseling for couples with fertility problems who seek therapy with assisted reproductive technologies such as *in vitro* fertilization.[64]

In the general population, miscarriages occur in about 15% of clinically-recognized pregnancies. About 3 to 5% of infants have a birth defect, and 1 to 3% have very low birth weight (< 1,500 g or 3 lb, 5 oz). Counseling about potential exposures can include statistics such as these to emphasize that a couple may experience reproductive problems through no fault of their own.

Among the most toxic pesticides to address during patient management are the organochlorine insecticides. In developed countries use of these agents has been curtailed, but some agents in this class remain in widespread use. Low doses of organochlorines are ubiquitous in food (especially fish) and are stored in the body. Very high exposures to toxic pesticides banned in the West are still occurring in the most impoverished countries around the globe. Poor farmers store pesticides in bedrooms or kitchens to prevent theft.[31] Women and men spray pesticides from backpacks without protective gloves, clothing or masks. Hot and humid weather makes safety equipment difficult to tolerate, but even more often it is neither available nor affordable.[31]

Most Americans get more pesticide exposure indoors than outdoors, especially in low-income housing.[44] Patients may believe that if a pesticide is widely available without restriction to consumers, or is applied professionally to their home, this must mean the product is safe. Having experienced no acute symptoms when using it in the past, patients may be surprised to learn that the agent is capable of increasing their risk for reproductive problems. Because the exposure dose makes a big difference, physicians can help patients estimate the likelihood of internal body exposure, and take precautions to lower exposures. The history can provide a rough estimate of exposure level from the quantity of pesticides applied, use of protective equipment, and likelihood of contaminating food, cigarettes or the home environment from pesticide residues on skin or clothing. Laboratory tests can be used to measure the level of exposure to certain pesticides a patient has recently used.[1] Detectable levels do not always equate with risk of reproductive disorders. Laboratory testing is most useful in the preconception period to help the patient improve pesticide handling practices so as to minimize internal body exposure.

CONCLUSION

The evidence available today shows that both men and women can experience adverse reproductive effects from pesticides. Every class of pesticides has at least one agent capable of causing a reproductive or developmental problem in laboratory animals or people. The animal studies support the biologic plausibility of a cause-effect relationship in humans for certain pesticides when the exposure dose is sufficiently large. Concurrence among multiple epidemiologic studies that controlled for confounding supports this conclusion, especially since epidemiologic research represents real-world exposures. The scientific advantage of population-based studies is often limited by imprecise exposure measurement, however, emphasizing the need for using more precise measures of absorbed dose in future epidemiologic

REFERENCES

1. Reigart JR, Roberts JR. Recognition and Management of Pesticide Poisonings. Washington DC: Office of Pesticide Programs, US Environmental Protection Agency; 1999. Available from: *http://www.epa.gov. oppfead1/safety/healthcare/handbook/handbook.pdf* [cited 2006 Jun 22]

2. Recio R, Robbins WA, Borja-Aburto V, Moran-Martinez J, Froines JR, Hernandez RM, Cebrian ME. Organophosphorous pesticide exposure increases the frequency of sperm sex null aneuploidy. Environ Health Perspect. 2001 Dec;109(12):1237-40.

3. Padungtod C, Hassold TJ, Millie E, Ryan LM, Savitz DA, Christiani DC, Xu X. Sperm aneuploidy among Chinese pesticide factory workers: scoring by the FISH method. Am J Ind Med. 1999 Aug;36(2) 230-8.

4. Sanchez-Pena LC, Reyes BE, Lopez-Carrillo L, Recio R, Moran-Martinez J, Cebrian ME, Quintanilla-Vega B. Organophosphorous pesticide exposure alters sperm chromatin structure in Mexican agricultural workers. Toxicol Appl Pharmacol. 2004 Apr 1;196(1):108-13.

5. Meeker JD, Singh NP, Ryan L, Duty SM, Barr DB, Herrick RF, Bennett DH, Hauser R. Urinary levels of insecticide metabolites and DNA damage in human sperm. Hum Reprod. 2004 Nov;19(11):2573-80.

6. Pina-Guzman B, Solis-Heredia MJ, Quintanilla-Vega B. Diazinon alters sperm chromatin structure in mice by phosphorylating nuclear protamines. Toxicol Appl Pharmacol. 2005 Jan 15;202(2):189-98.

7. Xia Y, Cheng S, Bian Q, Xu L, Collins MD, Chang HC, Song L, Liu J, Wang S, Wang X. Genotoxic effects on spermatozoa of carbaryl-exposed workers. Toxicol Sci. 2005 May;85(1):615-23.

8. Chauhan LK, Pant N, Gupta SK, Srivastava SP. Induction of chromosome aberrations, micronucleus formation and sperm abnormalities in mouse following carbofuran exposure. Mutat Res. 2000 Feb 16;465(1-2) 123-9.

9. Xia Y, Bian Q, Xu L, Cheng S, Song L, Liu J, Wu W, Wang S, Wang X. Genotoxic effects on human spermatozoa among pesticide factory workers exposed to fenvalerate. Toxicology. 2004 Oct 15;203(1-3):49-60.

10. Bian Q, Xu LC, Wang SL, Xia YK, Tan LF, Chen JF, Song L, Chang HC, Wang XR. Study on the relation between occupational fenvalerate exposure and spermatozoa DNA damage of pesticide factory workers. Occup Environ Med. 2004 Dec;61(12):999-1005.

11. Anway MD, Cupp AS, Uzumcu M, Skinner MK. Epigenetic transgenerational actions of endocrine disruptors and male fertility. Science. 2005 Jun 3 308(5727):1466-9.

testicular cell types from rats. Reprod Toxicol. 1995 Sep-Oct;9(5):461-73.

13. Ballering LA, Nivard MJ, Vogel EW A deficiency for nucleotide excision repair strongly potentiates the mutagenic effectiveness of methyl bromide in Drosophila. Mutagenesis. 1994 Jul;9(4):387-9.

14. Rignell-Hydbom A, Rylander L, Giwercman A, Jonsson BA, Lindh C, Eleuteri P, Rescia M, Leter G, Cordelli E, Spano M, Hagmar L. Exposure to PCBs and p,p'-DDE and human sperm chromatin integrity. Environ Health Perspect. 2005 Feb;113(2):175-9.

15. Traina ME, Rescia M, Urbani E, Mantovani A, Macri C, Ricciardi C, Stazi AV, Fazzi P, Cordelli E, Eleuteri P, Leter G, Spano M. Long-lasting effects of lindane on mouse spermatogenesis induced by in utero exposure. Reprod Toxicol. 2003 Jan-Feb;17(1):25-35.

16. Giri S, Giri A, Sharma GD, Prasad SB. Mutagenic effects of carbosulfan, a carbamate pesticide. Mutat Res. 2002 Aug 26;519(1-2):75-82.

17. Elbetieha A, Da'as SI, Khamas W, Darmani H. Evaluation of the toxic potentials of cypermethrin pesticide on some reproductive and fertility parameters in the male rats. Arch Environ Contam Toxicol. 2001 Nov; 41(4):522-8.

18. Yousef MI, El-Demerdash FM, Al-Salhen KS. Protective role of isoflavones against the toxic effect of cypermethrin on semen quality and testosterone levels of rabbits. J Environ Sci Health B. 2003 Jul;38(4): 463-78.

19. Rios AC, Salvadori DM, Oliveira SV, Ribeiro LR. The action of the herbicide paraquat on somatic and germ cells of mice. Mutat Res. 1995 Apr;328(1):113-8.

20. Eaton M, Schenker M, Whorton MD, Samuels S, Perkins C, Overstreet J. Seven-year follow-up of workers exposed to 1,2-dibromo-3-chloropropane. J Occup Med. 1986 Nov;28(11):1145-50.

21. Potashnik G, Porath A. Dibromochloropropane (DBCP): a 17-year reassessment of testicular function and reproductive performance. J Occup Environ Med. 1995 Nov;37(11):1287-92.

22. Quandt SA, Doran AM, Rao P, Hoppin JA, Snively BM, Arcury TA. Reporting pesticide assessment results to farmworker families: development, implementation, and evaluation of a risk communication strategy. Environ Health Perspect. 2004 Apr;112(5): 636-42.

23. Sallmen M, Liesivuori J, Taskinen H, Lindbohm ML, Anttila A, Aalto L, Hemminki K. Time to pregnancy among the wives of Finnish greenhouse workers. Scand J Work Environ Health. 2003 Apr;29(2):85-93.

24. Garry VF, Harkins M, Lyubimov A, Erickson L, Long L. Reproductive outcomes in the women of the Red River Valley of the north. I. The spouses of pesticide applicators: pregnancy loss, age at menarche, and exposures to pesticides. J Toxicol Environ Health A. 2002 Jun 14;65(11):769-86.

25. Garry VF, Harkins ME, Erickson LL, Long-Simpson LK, Holland SE, Burroughs BL. Birth defects, season of conception, and sex of children born to pesticide applicators living in the Red River Valley of Minnesota, USA. Environ Health Perspect. 2002 Jun;110 Suppl 3:441-9.

26. Arbuckle TE, Lin Z, Mery LS. An exploratory analysis of the effect of pesticide exposure on the risk of spontaneous abortion in an Ontario farm population. Environ Health Perspect. 2001 Aug;109(8):851-7.

27. Arbuckle TE, Savitz DA, Mery LS, Curtis KM. Exposure to phenoxy herbicides and the risk of spontaneous abortion. Epidemiology. 1999 Nov;10(6):752-60.

28. Nordby KC, Andersen A, Irgens LM, Kristensen P. Indicators of mancozeb exposure in relation to thyroid cancer and neural tube defects in farmers' families. Scand J Work Environ Health. 2005 Apr;31(2):89-96.

29. Garcia AM, Benavides FG, Fletcher T, Orts E. Paternal exposure to pesticides and congenital malformations. Scand J Work Environ Health. 1998 Dec;24(6): 473-80.

30. Farr SL, Cooper GS, Cai J, Savitz DA, Sandler DP. Pesticide use and menstrual cycle characteristics among premenopausal women in the Agricultural Health Study. Am J Epidemiol. 2004 Dec 15;160(12):1194-204.

31. Kishi M, Ladou J. International pesticide use. Introduction. Int J Occup Environ Health. 2001 Oct-Dec; 7(4):259-65.

32. Greenlee AR, Arbuckle TE, Chyou PH. Risk factors for female infertility in an agricultural region. Epidemiology. 2003 Jul;14(4):429-36.

33. Greenlee AR, Ellis TM, Berg RL. Low-dose agrochemicals and lawn-care pesticides induce developmental toxicity in murine preimplantation embryos. Environ Health Perspect. 2004 May;112(6):703-9.

34. Heindel JJ, Chapin RE, Gulati DK, George JD, Price CJ, Marr MC, Myers CB, Barnes LH, Fail PA, Grizzle TB, Schwetz BA, Yang RSH. Assessment of the reproductive and developmental toxicity of pesticide/ fertilizer mixtures based on confirmed pesticide contamination in California and Iowa groundwater. Fundam Appl Toxicol. 1994 May;22(4):605-21.

35. Stoker TE, Jeffay SC, Zucker RM, Cooper RL, Perreault SD. Abnormal fertilization is responsible for reduced fecundity following thiram-induced ovulatory delay in the rat. Biol Reprod. 2003 Jun;68(6):2142-9.

36. Eskenazi B, Mocarelli P, Warner M, Samuels S, Vercellini P, Olive D, Needham LL, Patterson DG Jr, Brambilla P, Gavoni N, Casalini S, Panazza S, Turner W, Gerthoux PM. Serum dioxin concentrations and endometriosis: a cohort study in Seveso, Italy. Environ Health Perspect. 2002 Jul;110(7):629-34.

37. Narotsky MG, Best DS, Guidici DL, Cooper RL. Strain comparisons of atrazine-induced pregnancy loss in the rat. Reprod Toxicol. 2001 Jan-Feb;15(1):61-9.

38. Thomas DC, Petitti DB, Goldhaber M, Swan SH, Rappaport EB, Hertz-Picciotto I. Reproductive outcomes

in relation to malathion spraying in the San Francisco Bay Area, 1981-1982. Epidemiology. 1992 Jan;3(1):32-9.

39. Longnecker MP, Klebanoff MA, Dunson DB, Guo X, Chen Z, Zhou H, Brock JW. Maternal serum level of the DDT metabolite DDE in relation to fetal loss in previous pregnancies. Environ Res. 2005 Feb;97(2): 127-33.

40. Garry VF, Schreinemachers D, Harkins ME, Griffith J. Pesticide appliers, biocides, and birth defects in rural Minnesota. Environ Health Perspect. 1996 Apr; 104(4):394-9.

41. Engel LS, O'Meara ES, Schwartz SM. Maternal occupation in agriculture and risk of limb defects in Washington State, 1980-1993. Scand J Work Environ Health. 2000 Jun;26(3):193-8.

42. Loffredo CA, Silbergeld EK, Ferencz C, Zhang J. Association of transposition of the great arteries in infants with maternal exposures to herbicides and rodenticides. Am J Epidemiol. 2001 Mar 15;153(6): 529-36.

43. Kristensen P, Irgens LM, Andersen A, Bye AS, Sundheim L. Birth defects among offspring of Norwegian farmers, 1967-1991. Epidemiology. 1997 Sep;8(5): 537-44.

44. Whyatt RM, Camann D, Perera FP, Rauh VA, Tang D, Kinney PL, Garfinkel R, Andrews H, Hoepner L, Barr DB. Biomarkers in assessing residential insecticide exposures during pregnancy and effects on fetal growth. Toxicol Appl Pharmacol. 2005 Aug 7;206(2): 246-54.

45. Eskenazi B, Harley K, Bradman A, Weltzien E, Jewell NP, Barr DB, Furlong CE, Holland NT. Association of in utero organophosphate pesticide exposure and fetal growth and length of gestation in an agricultural population. Environ Health Perspect. 2004 Jul;112(10): 1116-24.

46. Munger R, Isacson P, Hu S, Burns T, Hanson J, Lynch CF, Cherryholmes K, Van Dorpe P, Hausler WJ Jr. Intrauterine growth retardation in Iowa communities with herbicide-contaminated drinking water supplies. Environ Health Perspect. 1997 Mar;105(3):308-14.

47. Siddiqui MK, Srivastava S, Srivastava SP, Mehrotra PK, Mathur N, Tandon I. Persistent chlorinated pesticides and intra-uterine foetal growth retardation: a possible association. Int Arch Occup Environ Health. 2003 Feb;76(1):75-80.

48. Campoy C, Jimenez M, Olea-Serrano MF, Moreno-Frias M, Canabate F, Olea N, Bayes R, Molina-Font JA. Analysis of organochlorine pesticides in human milk: preliminary results. Early Hum Dev. 2001 Nov;65 Suppl:S183-90.

49. Smith D. Worldwide trends in DDT levels in human breast milk. Int J Epidemiol. 1999 Apr;28(2):179-88.

50. Ribas-Fito N, Grimalt JO, Marco E, Sala M, Mazon C, Sunyer J. Breastfeeding and concentrations of HCB and p,p'-DDE at the age of 1 year. Environ Res. 2005 May;98(1):8-13.

51. Slotkin TA. Cholinergic systems in brain development and disruption by neurotoxicants: nicotine, environmental tobacco smoke, organophosphates. Toxicol Appl Pharmacol. 2004 Jul 15;198(2):132-51.

52. McGready R, Hamilton KA, Simpson JA, Cho T, Luxemburger C, Edwards R, Looareesuwan S, White NJ, Nosten F, Lindsay SW. Safety of the insect repellent N,N-diethyl-M-toluamide (DEET) in pregnancy. Am J Trop Med Hyg. 2001 Oct;65(4):285-9.

53. McGeady SJ. Immunocompetence and allergy. Pediatrics. 2004 Apr;113(4 Suppl):1107-13.

54. Bilrha H, Roy R, Moreau B, Belles-Isles M, Dewailly E, Ayotte P. In vitro activation of cord blood mononuclear cells and cytokine production in a remote coastal population exposed to organochlorines and methyl mercury. Environ Health Perspect. 2003 Dec;111(16): 1952-7.

55. Weisglas-Kuperus N, Vreugdenhil HJ, Mulder PG. Immunological effects of environmental exposure to polychlorinated biphenyls and dioxins in Dutch school children. Toxicol Lett. 2004 Apr 1;149(1-3):281-5.

56. Rooney AA, Matulka RA, Luebke RW. Developmental atrazine exposure suppresses immune function in male, but not female Sprague-Dawley rats. Toxicol Sci. 2003 Dec;76(2):366-75.

57. Kariuki SK, ter Kuile FO, Wannemuehler K, Terlouw DJ, Kolczak MS, Hawley WA, Phillips-Howard PA, Orago AS, Nahlen BL, Lal AA, Shi YP. Effects of permethrin-treated bed nets on immunity to malaria in western Kenya I. Antibody responses in pregnant women and cord blood in an area of intense malaria transmission. Am J Trop Med Hyg. 2003 Apr;68(4 Suppl):61-7.

58. Ma X, Buffler PA, Gunier RB, Dahl G, Smith MT, Reinier K, Reynolds P. Critical windows of exposure to household pesticides and risk of childhood leukemia. Environ Health Perspect. 2002 Sep;110(9):955-60.

59. Alexander FE, Patheal SL, Biondi A, Brandalise S, Cabrera ME, Chan LC, Chen Z, Cimino G, Cordoba JC, Gu LJ, Hussein H, Ishii E, Kamel AM, Labra S, Magalhaes IQ, Mizutani S, Petridou E, de Oliveira MP, Yuen P, Wiemels JL, Greaves MF. Transplacental chemical exposure and risk of infant leukemia with MLL gene fusion. Cancer Res. 2001 Mar 15;61(6):2542-6.

60. Borkhardt A, Wilda M, Fuchs U, Gortner L, Reiss I. Congenital leukaemia after heavy abuse of permethrin during pregnancy. Arch Dis Child Fetal Neonatal Ed. 2003 Sep;88(5):F436-7.

61. Von Busch TA, Frazier LM, Sigler SJ, Molgaard CA. Feasibility of maternity protection in early pregnancy. Int J Occup Environ Health. 2002 Oct-Dec;8(4): 328-31.

62. Johnson K, Posner SF, Biermann J, Cordero JF, Atrash HK, Parker CS, Boulet S, Curtis MG; CDC/ ATSDR Preconception Care Work Group; Select Panel on Preconception Care. Recommendations to improve preconception health and health care—United States. A

report of the CDC/ATSDR Preconception Care Work Group and the Select Panel on Preconception Care. MMWR Recomm Rep 2006;55(RR-6):1-23.

63. Posner SF, Johnson K, Parker C, Atrash H, Biermann J. The national summit on preconception care: a summary of concepts and recommendations. Matern Child Health J 2006;10 Suppl 7:199-207.

64. Grainger DA, Frazier LM, Rowland CA. Preconception care and treatment with assisted reproductive technologies. Matern Child Health J 2006;10 Suppl 7:161-4.

doi:10.1300/J096v12n01_04

Carcinogenicity of Agricultural Pesticides in Adults and Children

Michael C. R. Alavanja, DrPH
Mary H. Ward, PhD
Peggy Reynolds, PhD

SUMMARY. The role of specific agricultural pesticides in relation to adult and childhood cancers has not been firmly established due to the lack of precise exposure data in previous studies. Improvements in exposure assessment, disease classification, and application of molecular techniques in recent epidemiological evaluations is rapidly improving our ability to evaluate the human carcinogenicity of agricultural pesticides. The role of pesticide exposures in the etiology of human cancer is outlined by anatomical site and recent development in exposure assessment and molecular epidemiology are summarized and evaluated. doi:10.1300/J096v12n01_05 *[Article copies available for a fee from The Haworth Document Delivery Service: 1-800-HAWORTH. E-mail address: <docdelivery@haworthpress.com> Website: <http://www.HaworthPress.com>]*

KEYWORDS. Pesticides, occupational exposures, environmental exposures, childhood cancer, cancer incidence

INTRODUCTION

Substances used to destroy, repel, or mitigate pests ranging from insects, and weeds to microorganisms are broadly labeled pesticides.[1] Currently human carcinogenicity risk assessment of pesticides is based largely on the results of experimental studies conducted on animals.[1] Although more reliable human risk assessments could come from human metabolism, mode of action, and/or epidemiological studies of pesticides with reliable information on exposure and potential confounders, historically few studies were available for use by regulatory or health agencies. The adequacy of using animal surrogates to assess human risk is questionable because toxicologic testing usually over-simplifies biologic reality, in that a single active ingredient is administered to genetically inbred strains of laboratory animals over a relatively short period of time. In contrast, the human population is genetically and

Michael C. R. Alavanja and Mary H. Ward are affiliated with the Division of Cancer Epidemiology and Genetics, National Cancer Institute, Rockville, MD.

Peggy Reynolds is affiliated with the California Department of Health Services, Environmental Health Investigations Branch.

Address correspondence to: Michael C. R. Alavanja, Captain USPHS, Division of Cancer Epidemiology and Genetics, National Cancer Institute, 8120 Executive Boulevard, Room 8000, Rockville, MD 20892 (E-mail: alavanjm@mail.nih.gov).

[Haworth co-indexing entry note]: "Carcinogenicity of Agricultural Pesticides in Adults and Children." Alavanja, Michael C. R., Mary H. Ward, and Peggy Reynolds. Co-published simultaneously in *Journal of Agromedicine* (The Haworth Medical Press, an imprint of The Haworth Press, Inc.) Vol. 12, No. 1, 2007, pp. 39-56; and: *Proceedings from the Medical Workshop on Pesticide-Related Illnesses from the International Conference on Pesticide Exposure and Health* (ed: Ana Maria Osorio, and Lynn R. Goldman) The Haworth Medical Press, an imprint of The Haworth Press, Inc., 2007, pp. 39-56. Single or multiple copies of this article are available for a fee from The Haworth Document Delivery Service [1-800-HAWORTH, 9:00 a.m. - 5:00 p.m. (EST). E-mail address: docdelivery@haworthpress.com].

Available online at http://ja.haworthpress.com
doi:10.1300/J096v12n01_05

physiologically diverse, exposures are typically to complex mixtures and occur to a broad age segment of the population, exposures can take place over decades, and either genotoxic or epi-genetic mechanisms may be at work. As a result, the human carcinogenicity data for almost all pesticides on the market today is lacking sufficient scientific evaluation. Under these circumstances, it is also difficult to establishing strong a priori hypotheses for epidemiological studies evaluating the potential link between specific pesticide and cancer(s). Without credible a priori hypotheses, scientific consensus-building regarding the interpretation of individual epidemiologic studies may be complicated and contentious.

While toxicologic testing of laboratory animals has undoubtedly kept many dangerous chemicals destined for agricultural, household and other commercial use off the market, recent experience from mass marketed pharmaceuticals (e.g., celebrex) have dramatically illustrated that even extensive pre-market laboratory testing is not foolproof.[2] Some dangerous chemicals do make it to the market place. It is informative to note that, currently, the International Agency for Research on Cancer has classified "occupational exposures in spraying and application on non-arsenical insecticides" as a group as "probable human carcinogens," yet only arsenical insecticides and TCDD (a contaminant of the phenoxy herbicide 2,4,5-T) have been identified as proven human carcinogens.[3,4] This is a precarious public health situation, since hundreds of millions of people are exposed to pesticides world-wide as either a direct result of occupational exposure in agricultural, public health or commercial use, or as a result of indirect by-stander exposures. In 1999, over 1 billion pounds of pesticides were applied in the United States and over 5.6 billion pounds were applied world wide.[5]

In vitro studies of human metabolism of pesticides has just begun to delineate the metabolic pathway of a number of pesticides, including chlorpyrifos, carbaryl, fipronyl, carbofuran and others.[6-17] The mode of action of human carcinogens includes a cascade of events that is initiated by exposure to the carcinogen and culminates in the manifestation of a cancer. Even in an early step of this process, pesticide metabolism has been found to be innately different in the human population resulting from genetic polymorphisms in the exogenous metabolic enzymes (XMEs) and in the physiologic status of an individual determined by age, diet, and exposure history.

Other sources of information concerning pesticide carcinogenicity include the U.S. Environmental Protection Agency's Office of Pesticide Programs list of chemicals evaluated for carcinogenic potential that can be reached at: http://npic.orst.edu/chemicals_evaluated_July2004. pdf and a more extensive list produced by the same office entitled "Reference Dose Tracking Report" which can be reached at: http://npic.orst.edu/tracking.htm

EXPOSURE ASSESSMENT

Recent biomarker studies indicate that exposure to pesticides is widespread in both adults and children in the U.S.[18-22] Direct exposures occur to individuals who personally apply pesticides in agricultural, occupational, or residential settings, and are likely to result in the highest levels of exposure. In contrast, environmental exposures are generally lower, but they are much more common and can occur more frequently than occupational exposures, with different and varied routes of exposure.[23]

Historically, the epidemiologic evaluation of individual pesticides for human carcinogenicity has been hampered by inadequate exposure assessment. The first generation of occupational epidemiological studies has inferred pesticide exposure from occupational or industrial classifications of work histories.[24,25] In the case of farm occupations, pesticides exposures have sometimes been inferred from knowledge of the crops grown in a particular area in a particular year.[26] Recognizing the weaknesses of these earlier studies, a few more recent occupational studies have expended greater resources in ascertainment of information about pesticide exposures directly and comprehensively and have validated these estimates using pesticide measurements in the field.[27] These exposure assessment efforts include job exposure matrices and exposure algorithms for occupational settings.

Exposure assessment is most difficult when attempting to determine if particular pesticides are associated with cancer in the general environment.[28-30] The general population is exposed to agricultural pesticides through drinking water, air, dust, and food. The U.S. Department of Agriculture estimates that 50 million people in the United States obtain their drinking water from groundwater that is potentially contaminated by pesticides and other agricultural chemicals.[31] Herbicides are detected most frequently, and a 1995 survey by an environmental organization found widespread contamination of tap water by herbicides, frequently at levels exceeding the U.S. EPA lifetime health advisory level.[32] Food can be contaminated by pesticides, particularly insecticides.[33,34] Although diet does not appear to be a major route of exposure for most pesticides,[35] there are concerns about the effects on children, who may be particularly sensitive. Infants can also be exposed to pesticides and pesticide metabolites in breast milk and via placental transfer.[36,37]

Agricultural pesticide use near populated areas is increasing nationwide with the increasing population growth into agricultural land.[38] The general population in agricultural areas, especially children, has been the focus of a number of studies to assess exposure levels and determinants. Most studies measured organophosphate insecticides[22,39-42] but a recent study measured corn and soybean herbicides.[40] In homes with pesticide applicators, children's exposure and residential pesticides levels are higher than homes without pesticide applicators and homes in non-farming areas.[22,39-41] Children in agricultural communities have exposure throughout the year but exposures are higher during the pesticide spraying season.[41] A recent study by Bradman et al.[42] found that pregnant women living in an agricultural area had 2.5 times higher urinary metabolite levels of OP pesticides compared to the general United States population. Potential pathways for exposure include pesticide drift, pesticide-laden dust tracked into the homes on shoes and on pets, and occupational take-home exposure on shoes and clothing with the latter pathway being particularly important.[20,39,40] Within the home, Gurunathan et al., observed that children from 3-6 years old received a negligible portion of their exposure from inhalation pathways, while dermal and non-dietary oral doses from playing with toys contributed the largest portion of their dose.[43] Repeated applications of pesticides could lead to continued accumulation in toys and other absorbent surfaces, e.g., pillows, with large absorbent reservoirs, which can become a long-term source of exposure for children. Similar results were observed by Sexton et al in a poor inner-city environment.[44] In Quebec, Valcke et al. observed that dietary ingestion of organophosphate pesticides were the main source of childhood exposure, suggesting that Quebec food may contain higher residue levels than elsewhere and more fruits and vegetables may be consumed.[45]

The methods used to assess general population exposures to pesticides have typically included questionnaires, biomonitoring, and environmental monitoring. Biological monitoring represents the "gold standard" in exposure assessment, providing an integrated measurement of exposure from all pathways and routes; however, except for the more persistent pesticides, a biologic measure usually reflects recent exposures. Carpet dust is a useful medium for environmental monitoring because pesticides and other chemicals are protected from degradation. Pesticide levels in carpet dust samples are 10- to 200-fold higher than levels in air inside the home and in soil around the home.[46] A recent study found few indoor air samples with detectable herbicides; whereas, detections were common in dust.[40] Measurement of pesticides in carpet dust samples has been used in a few epidemiologic studies of cancer.[47] In one study,[47] pesticide residues in dust showed good correlations with self-reported home and garden use, and non-Hodgkin's lymphoma risk was positively associated with both termite treatments and chlordane levels in dust. The authors concluded that both exposure assessment methods are important because data on pesticide levels in carpet dust provides an objective measure of exposure, and questionnaires are the only way to assess pesticide exposures before current carpets were in place.[47]

Although the combined use of questionnaires and environmental monitoring appear promising for assessing exposure to home and garden pesticides, questionnaires are not useful for determining indirect exposure to agricultural pesticides because agricultural residents are unlikely to know the type and extent of pesticide use near their home. Crop-specific information on pesticide use is available through the California Pesticide Use Reporting database, the most comprehensive mandatory. pesticide reporting database in the United States. Since 1990, the database includes information on the amount of agricultural pesticides applied, the date and method of application, and the crop species and acres treated.[48] The reporting unit is one section of the Public Land Survey System (approximately one square mile). Using a GIS and this database,[49] several cancer epidemiology studies have assessed potential exposure to agricultural exposures at a relatively small scale.[28,29,50] However, this approach has limitations because substantial pesticide drift occurs at distances of less than a mile.[51]

Using remote sensed data (e.g., satellite imagery and aerial photography) and a geographic information system (GIS), methods have been developed to create historical crop maps for identifying residences with probable exposure from agricultural drift.[52-55] These approaches use land cover data sets to identify the location of crop fields near residences. Crop-specific pesticide use data is then linked to crop maps to create a metric that estimates the potential exposure for each residence. These methods require validation for individual pesticides. However, several studies have shown a relationship between proximity of homes to crop fields and pesticide levels in house dust or pesticide metabolites in urine.[56-58] Thus, the use of crop maps and other data coupled with mandatory pesticide use records may provide a useful approach to studying the health effects of residential exposure to agricultural pesticides.

In this review we will evaluate the strengths and limitations of current epidemiological evidence linking agricultural pesticides and cancer in adults and children, and current scientific evidence on biologic mode of action that need to be investigated to more fully assess the plausibility of emerging epidemiological findings.

Prostate Cancer

Prostate cancer is the most common malignancy among men in the United States and in most Western countries.[59] Age, family history of prostate cancer, African-American ethnicity, hormonal factors, and possibly a high consumption of animal fat and red meat are the most consistent risk factors reported.[60] The literature suggests that prostate cancer may also be elevated among farmers.[61,62] Potential risk factors for prostate cancer found on the farm include insecticides, fertilizers, herbicides, and other chemicals.[63-68] Consistent with those reports a large prospective study of pesticide applicators which included farmers and commercial applicators and the spouse of farmer applicators in the U.S.[69] showed an overall small but significant excess-risk of prostate cancer in the cohort, which was also demonstrated in both Iowa and North Carolina and both among farmers and commercial applicators. Age and family history of prostate cancer were the only two significant demographic risk factors for prostate cancer within the cohort as a whole. When these factors were controlled, use of chlorinated pesticides among applicators over 50 years of age and methyl bromide use were significantly associated with prostate cancer risk. Among those with a family history of prostate cancer several pesticides including butylate, a widely used herbicide and four commonly used organothiophosphate insecticides including coumaphos, fonofos, chlorpyrifos and phorate and a pyrethriod, permethrin all showed significant interactions odds ratios. These associations suggest, but do not prove, that a family history of prostate cancer may increase the susceptibility to the carcinogenic effects of these insecticides. Additional studies confirming these associations are need, along with molecular epidemiological studies to establish biological plausibility and to suggest a mode(s) of action.

Lung Cancer

Lung cancer is the leading cause of cancer death in the world and cigarette smoking is estimated to be responsible for over 85% of all cancer deaths.[70] As a group farmers smoke less than the rest of the population, resulting in significantly less lung cancer mortality.[26] Arsenical compounds have been causally linked to lung cancer.[4] Occupationally this link has been observed in vineyard workers exposed to arsenic based pesticides[71] and among arsenical pesticide manufacturers.[72] In rodent bioassays a variety of other pesticides have caused lung tumors, but the epidemiological evidence supporting such an association in humans has been inconclusive.[73] In a study of Florida pesticide applicators[74] and a follow-up study of the same population,[75] the risk of lung cancer rose with the number of years licensed reaching a two fold excess risk among those with more than 20 years of application experience. Barthel observed a lung cancer relative risk of 3.0 among East German agricultural workers who applied pesticides 20 or more years.[76] While specific pesticides responsible for this risk were not identified in either the Florida or the East German study organophosphate and carbamate insecticides and phenoxyacetic acid herbicides[76] were suspected causes. In a cohort study of four manufacturing plants in Germany[77] and in a pooled analysis of 36 cohorts from 12 countries phenoxy herbicides and/or contaminants (dioxins and furans) were associated with lung cancer mortality.[78] Other studies of pesticide applicators[79,80] and pesticide manufacturing workers,[81-83] however, did not confirm this excess risk of lung cancer. In the large Agricultural Health Study, two widely used herbicides, metolachlor and pendimethalin showed significant exposure-response association with lung cancer incidence, as did two widely used organophosphate insecticides chlorpyrifos and diazinon.[84,85]

Colorectal Cancer

Colorectal cancer is the third most common incident cancer in the United States, with over 106,000 new cases of colon cancer and over 40,000 rectal cancers cases diagnosed in the U.S. in 2004.[86] The substantial worldwide variation in incidence rates for these cancers may be explained by dietary and other environmental factors.[87] Although colorectal cancer is not commonly considered to have a large occupational component, risk of colorectal cancer have been reported in a number of industries and occupations.[88-93] Colorectal cancer is generally found to be decreased in farmers and has not usually been associated with pesticides exposure,[26,94] however a number of studies in the past 12 years have shown some positive associations. Increased risk of mortality from rectal cancer was observed in studies of farmers in Italy[95] and Iceland.[96] The risk of rectal cancer incidence was observed to increase three-fold

among pesticide manufacturing workers exposed to dieldrin and aldrin in The Netherlands.[97,98] Alachlor manufacturing workers in the United States were observed to be at excess risk of colorectal cancer[99,100] and colorectal cancer patients in Egypt were reported to have higher serum organochlorine levels than controls.[101] In a large prospective study of pesticide applicators which included farmers and commercial applicators and the spouse of farmer applicators in the U.S.,[94] the overall colorectal cancer incidence was significantly less than expected based on comparison to incidence rates in the states of Iowa and North Carolina, where the applicators lived. In the same cohort, however, a significant exposure response trend was observed for rectal cancer incidence among chorpyrifos users (p for trend = 0.008), rising to over 2.5 in the highest exposure category and a significant exposure response trend was observed for colon cancer incidence among aldicarb users (p for trend < 0.001).[102] While unexpected, these data provide the strongest suggestive evidence to date for an association between specific pesticide exposures and the incidence of colorectal cancer. Additional studies confirming these associations are needed along with molecular epidemiological studies to establish biological plausibility and to suggest a mode(s) of action.

Pancreatic Cancer

A number of occupational studies of agricultural workers and pesticide users show an elevated risk of pancreatic cancer including farmers,[103-105] licensed and unlicensed pesticide applicators,[106-110] lawn care workers[111] and workers that might be exposed to pesticides.[112] Significantly elevated risk of pancreatic cancer was observed among DDT manufacturing workers (OR = 4.8, 95% CI 1.3-17.6).[113] A study from Australia showed a similar approximately five-fold risk associated with DDT application.[114] In yet another study, long-term residents residences living in zip codes with the highest use of four pesticides including 1,3-dichloropropene (1,3-d), captafol, pentachloronitrobenzene (PCNB), and dieldrin had significantly higher pancreatic mortality compared to residents in other zip codes in three agriculturally important counties in California.[115] Other studies of pesticide manufacturing workers[116-121] and farmers and farm workers[122-124] did not shown an excess, however. Since pancreatic cancer is quickly fatal it is difficult to obtain exposure information through in-person interview, prospective studies offer a methodologic advantage to study diseases of high case fatality.

Non-Hodgkin's Lymphoma

Among the occupational studies of farmers, 18 of 29 studies, showed an excess of non-Hodgkin's lymphoma compared to the general population.[26] In a meta-analysis of 6 studies examining the association between non-Hodgkin's lymphoma and farming in the central United States the overall estimated relative risk is 1.34 (95% Confidence interval = 1.17-1.55).[125] The various studies which attempted to assess the associations between specific pesticides and non-Hodgkin's lymphoma, however, do not yet offer a clear etiologic picture. Among a series of population-based case-control studies, NHL was linked to organochlorine pesticides.[126-129] In a multi-center population-based incident study in Canada[127] several chemical classes of pesticides were associated with non-Hodgkin's lymphoma including phenoxy and benzoic acid herbicides and to carbamate and organophosphate insecticides, to amide fumigants and to the fumigant carbon tetrachloride. Carbaryl, a carbamate insecticide, was found to be a risk factor for NHL in a pooled analysis.[130] A trace contaminant, dioxin, formed during the manufacture of several chemicals including the phenoxyherbicides 2,4,5 T and 2,4,5, TP has also been suggested to play a role in the etiology of non-Hodgkin's lymphoma.[131] Additional studies are needed that have the capability of identifying a association of a specific pesticide with NHL are need, since NHL is a grouping of more than 20 phenotypes, it may be necessary for future studies to distinguish between these phenotype in etiologic studies as well.[132]

Leukemia

A number of epidemiologic and environmental toxicology studies have provided evidence to support an etiologic association between insecticide exposures and leukemia. However, the variety of malignancies under the general title leukemia may have limited the etiologic picture linking pesticides to specific leukemias in the literature historically. Although the evidence supporting a link between any specific pesticide or agricultural exposure and leukemia is not strong, our greatest etiologic insights may come from studies which focus on specific malignancies.

Among a farming and animal breeding area of Northern Italy a case-control study of Chronic Lymphocytic Leukemia (CLL) was associated with farm-animal breeding. The authors suggest that carbamate, organophosphates, and DDT may be responsible.[133]

A significant association between organophosphate insecticide exposure and Hairy Cell Leukemia risk was observed in a hospital-based study in France.[134] While herbicides, not organophosphate insecticides, were associated with Hairy-Cell Leukemia and non-Hodgkin's Lymphoma in a pooled analyses of two case-control studies conducted in Sweden.[135]

The biological plausibility and mode of action linking pesticides and certain leukemias has been the focus of a number of studies in the past few years. A dose-dependent leukemic cell proliferation has been observed when increasing doses of isofenphos (an organophosphate insecticide) were administered to K562 leukemic cell lines.[136] While in vitro studies are not always replicated in whole animals or humans, exploration of the mode of action of these pesticides in vitro has begun to deepen our understanding of the pathobiology of the process and suggesting new approaches for molecular epidemiology in exposed human populations.[137,138]

Multiple Myeloma

A meta-analysis of 32 studies of multiple myeloma and farming, published between 1981-1996, estimated the relative risk among farmers to be similar 1.23 (95% CI 1.14-1.32).[139] Infectious agents, solvents, and pesticides have been proposed as etiologic agents but to date the evidence in support of any one of these agents is not strong.[1] Subsequent to the publication of this meta-analysis, at least two additional studies supporting a link between agricultural exposures and multiple myeloma have been reported.[140,141]

Soft-Tissue Sarcoma

Soft-tissue sarcomas are a heterogeneous mix of many subtypes. This diverse mixture of subtype is probably responsible for the inconsistent pattern of observations that have characterize the published literature to date. A number of positive associations between pesticide use and soft-tissue sarcoma have been published,[142-146] but a number of others have found no association.[147-149] Attempting to study the etiology of soft-tissue sarcoma as one entity, may mask meaningful associations with specific-subtypes. Hoppin et al. observed that herbicide use was significantly associated with malignant fibrohistiocytic sarcoma but not with liposarcoma.[150]

Other Cancers in Adults

Cancers of the female breast, ovaries, testicles, liver, kidney, brain and Hodgkin's disease have been associated with pesticide use in some studies, but the links are weak, and/or inconsistent, and the data are not sufficient to draw conclusions at this time. For example, the estrogenic activity of PCBs and DDE were thought to contribute to increased risk of breast cancer,[151-154] however, many newer studies including both cohort studies and case-control studies show no convincing association.[155-160] In a prospective study of pesticide applicators and their spouse, female applicators were at a significantly elevated risk (i.e., 8 observed cases, vs. 1.9 expected) of ovarian cancer, but the female spouses were not at any increased risk.[161] Because there were only a small numbers of ovarian cancer cases, the etiologic agent responsible for this excess was not discovered. In meta-analyses of Hodgkin's disease a small excess risk was observed among farmers but an infectious etiology or another agricultural chemical etiology were equally plausible.[162] The literature for testicular, liver, kidney and brain cancer is even less informative and larger studies with substantially better exposure assessment are sorely needed.[163-166]

Childhood Cancers

A number of studies now suggest that household pesticide use may be associated with a higher risk of childhood cancers,[167-169] but few studies have systematically evaluated the risk relationships for exposures to agricultural pesticides. The very comprehensive reviews of Daniels[168] and Zahm[169] provide in depth discussions of the evidence for higher risks of childhood cancer with household pesticide use, particularly during the *in utero* period. Those reviews show scattered evidence for pesticide-associated risks both for the most common malignancies in children, the leukemias and brain tumors, but also for neuroblastoma, non-Hodgkin's lymphoma, Wilms' tumor, Ewings sarcoma, and germ cell tumors. This discussion will focus on agricultural pesticides.

The evidence for the agricultural exposures and childhood cancers has been based primarily on the identification of paternal or maternal occupations in farming, and/or self report of working with agricultural pesticides.[169,170] Because the prevalence of agricultural occupations is rare in population-based case-control studies, risk estimates have been generally unstable. Notably, however, a pooled analysis from a large international consortium of studies including over 1,200 cases of childhood brain tumors across seven countries suggested a rather consistent pattern of elevated risks associated with both animal husbandry and parental pesticide use.[171] While these studies have provided provocative leads, little detailed information on specific pesticides or workplace activities has been available. More recently a number of investigators have undertaken studies targeted at better understanding the contribution that agricultural pesticides, a very different group of agents than those used in household settings, might make to childhood cancer etiology. These studies have required the use of somewhat more creative strategies in study design and exposure assessment than the more traditional methods of the past. This sparse literature is summarized in Table 1.

The earliest studies come from Europe. A very interesting study from Norway examined the cancer experience of a large cohort born between 1952-1971 to parents identified as farm holders in the Norwegian agricultural censuses between 1969-1989, and linked to the Cancer Registry of Norway.[172] The agricultural censuses, mandatory and the basis for farm subsidies from the government, are a rich source of

TABLE 1. Studies of Agricultural Pesticide Exposure and Childhood Cancer

Case-Control Studies				
Author/Year/Place	*Cases*	*Controls*	*Exposure Measure(s)*	*Findings*
Meinert, 1996 Northern Germany	Acute leukemia cases from the German Childhood Cancer Registry: N = 161.	Selected from community resident rosters. N = 161	Use of pesticides on farms.	OR = 1.6 (1.0-6.1)
Meinert, 2000 West Germany	Mainz cancer registry and cases identified for another national study: Leukemia, N = 1,184 Lymphoma, N = 234 Solid tumors, N = 940	Selected from community resident rosters. Matched on gender, birth year and community. N = 2,5888	Parental occupational exposures to herbicides, insecticides, fungicides. (via questionnaire)	Father's exposures (ever): Leukemia OR = 1.6 (1.1-2.3) Lymphoma OR = 1.9 (0.9-3.7) Mother's exposures (ever): Leukemia OR = 2.5 (1.3-4.7) Lymphoma OR = 4.1 (1.1-16)
Reynolds, 2005 California, USA	Population-based cases (born in California) from the cancer registry aged 0-4, 1990-1997. All Sites, N = 2,189 Leukemia, N = 1,658 CNS tumors, N = 695	Selected from statewide birth files. Matched on birth date and sex. N = 4,335	Agricultural pesticide use in the 9 months before birth within a half mile of the maternal residence at birth.	No associations except for two commonly used pesticides (use greater than 50th percentile) and leukemia: Metam sodium OR = 2.05 (1.01-4.17) Dicofol OR = 1.83 (1.05-3.22)
Other Study Designs				
Author/Year/Place	*Cases*	*Subjects*	*Exposure Measure(s)*	*Findings*
Kristensen, 1996 Norway	Population-based from the Cancer Registry of Norway, ages 0-14, 1988-1994: N = 437	Offspring born 1952-1991 to parents identified as farm holders in the agricultural censuses. N = 323,292	Detail from the agricultural censuses including size of farms, farming activity, pesticide purchase.	Astrocytomas. Age 0-14 among offspring of pig farmers: OR = 3.37 (1.63-6.94) All cancers age 0-4 among offspring of owners of orchards or greenhouses: OR = 1.86 (1.18-2.92)
Reynolds, 2002 California, USA	Population-based California Cancer Registry cases ages 0-14, 1988-1994: All sites, N = 6,988 Leukemia, N = 2,443 Glioma, N = 1,351	Population data at the block group level	Average annual agricultural pesticide use (4 toxicologic groups, 4 chemical groups, 7 individual chemicals) at the block group.	No associations except for propargite (at 90th percentile of use, leukemia OR = 1.48, 95% CI = 1.03-2.13)
Flower, 2004 Iowa, USA	Cancers 1975-1988 from the Iowa Cancer Registry aged 0-19: N = 50 cancers	Children of parents in the Agricultural Health Study. N = 17,357 children	Parents' pesticide use: mixing, applying and several specific pesticides and pesticide groups.	No significant associations except for paternal aldrin use OR = 2.66 (1.08-6.59)

detailed information on farming practices. This detail, coupled with the ability to link cohort members by family and by cancer outcome using a unique personal identification number, resulted in the opportunity to examine a number of detailed risk associations by age group and by cancer type. The investigators observed a nearly two-fold risk for cancer among young children (under age 5) born to parents with orchards or greenhouses, although based on only 17 exposed cases. Among the offspring of pig farmers, older children (diagnosed age 5-19 years old) exhibited an elevated cancer risk (OR = 1.33, 95% CI = 1.03-1.71). The generally increased risk with horticulture among younger children, and with animal husbandry for older children, tended to hold up for the leukemias, brain tumors, and several other childhood malignancies.

Following up on the observation of elevated childhood leukemia and pesticide use on farms in small case-control study in northwestern Germany,[173] investigators conducted a large case-control study in a broader region of the country designed to address similar issues with greater power.[174] The questionnaire covered not only farm occupations, but smaller scale farming as commonly practiced in the region, as well as details on the use of pesticides/herbicides and cattle breeding. Childhood cancer risks were elevated for maternal and paternal reported occupational exposure to herbicides, insecticides and fungicides, and were most elevated for reported maternal use during pregnancy (OR = 11.8, 95% CI = 2.2-664). Although a well designed study with many strengths, this study was criticized for using traditional questionnaire methods lacking in the ability to better elucidate exposure to specific chemical agents.[175]

Two large California studies, one an ecologic study[28] and one a case-control study of young children,[29] were designed to evaluate the risk patterns for childhood cancer with special attention to specific pesticides and groups of pesticides. Built on the availability of detailed pesticide use information available in the State, and using a geographic information system, these studies were able to assess small scale exposure potential for residents living near treated fields. Although these studies provided initial assessments for specific agents, because of their scale and complexity they were limited in their ability to account for other risk factors and were only able to target limited time windows of potential exposure.

In the first analysis of cancer among the offspring of participants in the Agricultural Health Study, most associations among children were null but there were suggestive findings for increased cancer risk among children of parents of children using aldrin.[176] Although this study provided the opportunity to assess risk associations in the context of a well characterized and well defined cohort, there were too few cancers among children in participant families to provide robust risk estimates.

Two of the primary challenges in studying childhood exposure potential to agricultural pesticides are the need for new exposure assessment methods and the assessment of detailed chemical information. Added challenges include the assessment of important temporal windows of exposure and greater geographic or personal precision for exposure. Furthermore, because of a lack of specifically identified biologic mechanisms and the broad net cast by the studies to date, there has been no clear pattern of risk associations. Future efforts will need to attend to ways of better bringing together these elements to study rare health outcomes such as childhood cancer.

Biologic Plausibility

Evaluating the biologic plausibility of the epidemiologic associations linking particular pesticides with particular cancers in well-designed biomarker studies is an additional important step in the etiologic evaluation of potential human carcinogen. Studies evaluating the link between exposures and early biologic effect, chronic biologic effect, or to preclinical markers of disease would add valuable detail to our proposed causal association. An array of the most promising markers were evaluated in a workshop held in Research Triangle Park, North Carolina, on March 1-2, 2005,[177] these included biomarkers of early effect: the generation of reactive intermediated such as reactive oxygen species,[178] the formation of DNA adducts,[178] the perturbation of the immune system,[179-185] and the use of proteomics[186] and host susceptibility factors including paraoxonase 2 activity,[187] cytochrome P450 polymorphisms,[188] glutathione conjugation enzymes,[189] DNA repair enzymes[190] and chromosome aberrations.[191] Some of these markers are discussed briefly below.

Biomarkers of Early Effect

Reactive intermediates: The concept that the carcinogenicity of a compound was related to the production of chemicals that formed in the body was first developed in the 1930s.[178] The basic principal can be explained by propagation of free radicals which interact with target tissue. An important component of free radical chemistry is oxidative damage, resulting from the production of partially reduced oxygen species. While much of the basic chemistry involved in leading up to toxicity is now understood, a more thorough understanding of cell biology will be necessary to help us: (a) choose biomarkers for monitoring human populations exposed to pesticides and other potential hazardous compounds; (b) choose laboratory-based assays that can be used to predict toxic and carcinogenic effects of new chemicals.

Chromosome aberrations: Significant seasonal increases in chromosome aberrations and micronucleus frequency (in the range from 11 to 15 micronuclei/1000 binuclear lymphocytes) were associated with fungicide applications in agricultural important regions of Minnesota.[191] Since chromosome aberrations have been associated with increased cancer risk, extending these techniques to the large analytical studies many prove very important.[191]

Immunotoxicology: There is evidence that some xenobiotics, including a number of pesticides, can cause immune suppression and or cause or exacerbate allergic disease. Several widely used pesticides including carbaryl, carbofuran, dieldrin, chlorpyrifos, and diazinon[179] can suppress the antibody response. There is also some evidence that carbofuran, dieldrin, clorpyrifos, and diazinon supress T cell proliferation, and/or the DTH response (delayed type hypersensitivity). Also, human exposure to certain pesticide mixtures has been associated with increased incidence of otitis media in infants[179] and increased incidence of herpes zoster (shingles) in adults.[179] To assess the increased risk of cancer several different biomarkers of the immune system may be of benefit. Tumor types that occur with increased frequency in immune deficient mice and humans are virus-associated tumors. Epstein-Barr virus (EBV) has been implicated in the etiology of several different lymphoid and epithelial malignancies including both non-Hodgkin's and Hodgkin's lymphoma. Determining antibody titers in individuals exposed to pesticides before immunization against influenza or hepatitis A and B would provide valuable immunotoxicologic data. Additionally the collection of uncoagulated blood samples, peripheral blood mononuclear cells could be collected for use in assays for NK activity (natural killer cells) and lymphocytte proliferation. Finally, monitoring incidence of disease including the recurrence of herpes zoster (shingles) could certainly be informative.[179]

Proteomics: The study of protein expression patterns in urine or blood after exposure to a possible carcinogen may be a biomarker technology that could be used to evaluate the early effect of exposure to certain pesticides. On going studies of arsenic ingestion and bladder cancer have used proteomic technology and early results suggest the technology can be used to reliably distinguish between hose highly exposed and those lightly exposed to drinking water arsenic.[186]

Susceptibility Factors Modulating Cancer Risk from Pesticides

A number of enzymes systems including the paraoxonases[187] and the CYP superfamily[188] are important in the activation and deactivation of widely used insecticides. Genetic polymorphisms in these enzyme systems and natural variations in the serum concentrations of these enzymes may theoretically be responsible for putting some segment of the population at enhanced risk.

Paraoxonase 1 (PON1) is an HDL associated enzyme that catalyzes a number of different reactions including the hydrolysis of the toxic oxon metabolites of the insecticides diazinon and chlorpyrifos and possibly other organothiophosphates. Evidence for the physiologic importance of PON1 in modulating exposure to these two important insecticides comes from several different laboratory studies.[187]

The glutathione S-transferace (GSTs) are a multi-gene family that catalyze the conjugation of electrophilic substrates with endogenous antioxidant, glutathione. Numerous studies have demonstrated that GSTM1 homozygous null genotype is associated with modest (30-40%) increase in bladder and lung cancer risk, supporting the hypothesis that deficient GST activity might increase risk for cancer in carcinogen-exposed populations.[189] Relatively few studies have examined whether GST polymorphisms might be associated with increased adverse health risks from pesticides or other agricultural chemicals, but it remains possible that genetic polymorphisms in human GST enzymes could increase or decrease sensitivity to certain pesticides.[189]

Defects in repair of lesions resulting from environmental exposures are associated with increased cancer due to defective nucleotide expression repair genes (NER), e.g., sunlight induced skin cancer due to defective NER.[190] Similarly, defects in the DNA mismatch repair (MMR) pathway and the resulting production of microsatellites is now established as a biomarker for the loss of MMR activity in tumor cells.[190] Evaluation of these biomarkers has not yet been done in highly pesticide exposed human population but the need to do these studies is compelling. Evaluating the importance of these enzyme systems in modulating cancer risk in human populations may soon be possible in large epidemiological studies that have documented exposures carefully and have collected relevant tissue samples.[192]

CONCLUSIONS

Currently, the results from both bioassays and epidemiology have yet to convincingly demonstrate the carcinogenic potential of most pesticides suspected of carcinogenicity. Only arsenic-containing insecticides and TCDD (a contaminant of phenoxy herbicide 2,4,5 T) are recognized as known human carcinogens by the International Agency for Research on Cancer, although the agency classified "occupational exposures in spraying and application of non-arsenical insecticides" as a group as "probable human carcinogens" (category 2A). Inconclusive bioassays and incomplete biological evaluations make it difficult/impossible to formulate strong a priori hypotheses that could be tested in epidemiological studies. While weak exposure assessment methods and small study sizes have characterized much of the epidemiology studies conducted in the 1970s and 1980s. Nonetheless, many valuable lessons were learned from these earlier studies. Advances in toxicology, exposure assessment and epidemiology, particularly molecular epidemiology have greatly increased the opportunities for making great progress in the evaluation and identification of human carcinogens that may still exist in the agricultural chemical marketplace. The integrated application of these disciplines in carefully designed studies should make the next five to ten years, a period of unparalleled progress in understanding pesticide toxicology and carcinogenicity.

REFERENCES

1. Fenner-Crisp PE. Risk assessment and risk management: the regulatory process. In: Kreiger R, editor. Handbook of pesticide toxicology. 2nd edition. San Diego (CA): Academic Press; 2001. pp. 681-690.

2. *www.fda.gov/bbs/topics/answers/2002/ans*

3. International Agency for Research on Cancer (IARC). Monographs on the evaluation of carcinogenic risks to humans. Occupational exposures to insecticides and some pesticides. Volume 53. Lyon, France: IARC; 1991. pp. 535.

4. International Agency for Research on Cancer (IARC). Monographs on the evaluation of carcinogenic risks to humans. Suppl 7. Overall evaluations of carcinogenicity: an updating of IARC monographs Volumes 1 to 42. Lyon, France: IARC; 1987.

5. Donaldson D, Kiely T, Grube A. Pesticides industry sales and usage. 1998 and 1999 market estimates. Washington (DC): US Environmental Protection Agency. Report No. EPA-733-R-02-001. Available from: *http://www.epa.gov/oppbead1/pestsales/99pestsales/market_estimates1999.pdf* [cited 2006 May 22].

6. Tang J, Cao Y, Rose RL, Brimfield AA, Dai D, Goldstein JA, Hodgson E. Metabolism of chlorpyrifos by human cytochrome P450 isoforms and human, mouse, and rat liver microsomes. Drug Metab Dispos. 2001 Sep;29(9):1201-1204.

7. Tang J, Cao Y, Rose RL, Hodgson E. In vitro metabolism of carbaryl by human cytochrome P450 and its inhibition by chlorpyrifos. Chem Biol Interact. 2002 Oct 20;141(3):229-241.

8. Tang J, Amin Usmani K, Hodgson E, Rose RL. In vitro metabolism of fipronil by human and rat cytochrome P450 and its interactions with testosterone and diazepam. Chem Biol Interact. 2004 Apr 15;147(3):319-329.

9. Dai D, Tang J, Rose R, Hodgson E, Bienstock RJ, Mohrenweiser HW, Goldstein JA. Identification of variants of CYP3A4 and characterization of their abilities to metabolize testosterone and chlorpyrifos. J Pharmacol Exp Ther. 2001 Dec;299(3):825-831.

10. Usmani KA, Rose RL, Goldstein JA, Taylor WG, Brimfield AA, Hodgson E. In vitro human metabolism and interactions of repellent N,N-diethyl-m-toluamide. Drug Metab Dispos. 2002 Mar;30(3):289-294.

11. Usmani KA, Rose RL, Hodgson E. Inhibition and activation of the human liver microsomal and human cytochrome P450 3A4 metabolism of testosterone by deployment-related chemicals. Drug Metab Dispos. 2003 Apr;31(4):384-391.

12. Usmani KA, Karoly ED, Hodgson E, Rose RL. In vitro sulfoxidation of thioether compounds by human cytochrome P450 and flavin-containing monooxygenase isoforms with particular reference to the CYP2C subfamily. Drug Metab Dispos. 2004 Mar;32(3):333-339.

13. Hodgson E. In vitro human phase I metabolism of xenobiotics I: pesticides and related compounds used in agriculture and public health, May 2003. J Biochem Mol Toxicol. 2003;17(4):201-206.

14. Hodgson E, Cherrington N, Coleman SC, Liu S, Falls JG, Cao Y, Goldstein JE, Rose RL. Flavin-containing monooxygenase and cytochrome P450 mediated metabolism of pesticides: from mouse to human. Rev Toxicol. 1998;2:231-243.

15. Choi J, Rose RL, Hodgson E. In vitro human metabolism of permethrin: the role of human alcohol and aldehyde dehydrogenases. Pesticide Biochem Physiol. 2002;73:117-128.

16. Coleman S, Liu S, Linderman R, Hodgson E, Rose RL. In vitro metabolism of alachlor by human liver microsomes and human cytochrome P450 isoforms. Chem Biol Interact. 1999 Aug 30;122(1):27-39.

17. Coleman S, Linderman R, Hodgson E, Rose RL. Comparative metabolism of chloroacetamide herbicides and selected metabolites in human and rat liver micro-

somes. Environ Health Perspect. 2000 Dec;108(12): 1151-1157.

18. Adgate JL, Barr DB, Clayton CA, Eberly LE, Freeman NC, Lioy PJ, Needham LL, Pellizzari ED, Quackenboss JJ, Roy A, Sexton K. Measurement of children's exposure to pesticides: analysis of urinary metabolite levels in a probability-based sample. Environ Health Perspect. 2001 Jun;109(6):583-590.

19. Barr DB, Bravo R, Weerasekera G, Caltabiano LM, Whitehead RD Jr, Olsson AO, Caudill SP, Schober SE, Pirkle JL, Sampson EJ, Jackson RJ, Needham LL. Concentrations of dialkyl phosphate metabolites of organophosphorus pesticides in the U.S. population. Environ Health Perspect. 2004 Feb;112(2):186-200.

20. Fenske RA, Lu C, Curl CL, Shirai JH, Kissel JC. Biological monitoring to characterize organophosphorous pesticide exposure among children and workers: an analysis of recent studies in Washington State. Environ Health Perspect 2005;113(11):1651-1657.

21. Shalat SL, Donnelly KC, Freeman NC, Calvin JA, Ramesh S, Jimenez M, Black K, Coutinho C, Needham LL, Barr DB, Ramirez J. Nondietary ingestion of pesticides by children in an agricultural community on the US/Mexico border: preliminary results. J Expo Anal Environ Epidemiol. 2003 Jan;13(1):42-50.

22. O'Rourke MK, Lizardi PS, Rogan SP, Freeman NC, Aguirre A, Saint CG. Pesticide exposure and creatinine variation among young children. J Expo Anal Environ Epidemiol. 2000 Nov-Dec;10(6 Pt 2):672-681.

23. Needham LL, Ozkaynak H, Whyatt RM, Barr DB, Wang RY, Naeher L, Akland G, Bahadori T, Bradman A, Fortmann R, Liu LJ, Morandi M, O'Rourke MK, Thomas K, Quackenboss J, Ryan PB, Zartarian V. Exposure assessment in the National Children's Study: introduction. Environ Health Perspect. 2005 Aug;113(8): 1076-1082.

24. Cantor KP, Blair A, Everett G, Gibson R, Burmeister LF, Brown LM, Schuman L, Dick FR. Pesticides and other agricultural risk factors for non-Hodgkin's lymphoma among men in Iowa and Minnesota. Cancer Res. 1992 May 1;52(9):2447-2455.

25. Reif JS, Pearce N, Fraser J. 1989. Occupational risks for brain cancer: a New Zealand Cancer Registry-based study. J Occup Med. 1989 Oct;31(10):863-867.

26. Blair A, Zahm SH. Cancer among farmers. Occup Med. 1991 Jul-Sep;6(3):335-354.

27. Coble J, Arbuckle T, Lee W, Alavanja M, Dosemeci M. The validation of a pesticide exposure algorithm using biological monitoring results. J Occup Environ Hyg. 2005 Mar;2(3):194-201.

28. Reynolds P, Von Behren J, Gunier RB, Goldberg DE, Hertz A, Harnly ME. Childhood cancer and agricultural pesticide use: an ecologic study in California. Environ Health Perspect. 2002 Mar;110(3):319-324.

29. Reynolds P, Von Behren J, Gunier RB, Goldberg DE, Harnly M, Hertz A. Agricultural pesticide use and childhood cancer in California. Epidemiology. 2005 Jan;16(1):93-100.

30. Zahm SH, Ward MH. Pesticides and childhood cancer. Environ Health Perspect. 1998 Jun;106 Suppl 3:893-908.

31. Nielson EG, Lee LK. The magnitude and costs of groundwater contamination from agricultural chemicals: a national perspective. Agricultural Economics Rpt no 576. Washington: US Department of Agriculture; 1987.

32. Cohen B, Wiles R, Condon E. Executive summary. In: Weed killers by the glass: a citizen's tap water monitoring project in 29 cities. Washington: Environmental Working Group; 1995. pp. 1-5.

33. Nigg HN, Beier RC, Carter O, Chaisson C, Franklin C, Lavy T, Lewis RG, Lombardo P, McCarthy JF, Maddy KT, Moses M, Norris D, Peck C, Skinner K, Tardiff RG. Exposure to pesticides. In: Baker SR, Wilkinson CF, editors. Advances in modern environmental toxicology; the effects of pesticides on human health. Vol XVIII. Princeton (NJ): Princeton Scientific; 1990. pp. 35-130.

34. National Research Council. Pesticides in the diet of infants and children. Washington: National Academy Press; 1993. Available from: http://darwin.nap.edu/books/0309048753/html [cited 2006 May 23].

35. Immerman FW, Firestone MP. Non-occupational pesticide exposure study (NOPES). Summary Report. Rpt no RTI/136/01-03D. Research Triangle Park (NC): Research Triangle Institute; 1989.

36. Gladen BC, Rogan WJ. DDE and shortened duration of lactation in a northern Mexican town. Am J Public Health. 1995 Apr;85(4):504-508.

37. Rogan WJ, Gladen BC. Study of human lactation for effects of environmental contaminants: the North Carolina Breast Milk and Formula Project and some other ideas. Environ Health Perspect. 1985 May;60: 215-221.

38. Lee S, McLaughlin R, Harnly M, Gunier R, Kreutzer R. Community exposures to airborne agricultural pesticides in California: ranking of inhalation risks. Environ Health Perspect. 2002 Dec;110(12):1175-1184.

39. Fenske RA, Lu C, Barr D, Needham L. Children's exposure to chlorpyrifos and parathion in an agricultural community in central Washington State. Environ Health Perspect. 2002 May;110(5):549-553.

40. Curwin BD, Hein MJ, Sanderson WT, Nishioka MG, Reynolds SJ, Ward EM, Alavanja MC. Pesticide contamination inside farm and nonfarm homes. J Occup Environ Hyg. 2005 Jul;2(7):357-367.

41. Koch D, Lu C, Fisker-Andersen J, Jolley L, Fenske RA. Temporal association of children's pesticide exposure and agricultural spraying: report of a longitudinal biological monitoring study. Environ Health Perspect. 2002 Aug;110(8):829-833.

42. Bradman A, Eskenazi B, Barr DB, Bravo R, Castorina R, Chevrier J, Kogut K, Harnly ME, McKone TE. Organophosphate urinary metabolite levels during pregnancy and after delivery in women living in an agricultural community. Environ Health Perspect. 2005 Dec;113(12):1802-1807.

43. Gurunathan S, Robson M, Freeman N, Buckley B, Roy A, Meyer R, Bukowski J, Lioy PJ. Accumulation of chlorpyrifos on residential surfaces and toys accessible to children. Environ Health Perspect. 1998;106:9-16.

44. Sexton K, Adgate JL, Fredrickson AL, Ryan AD, Needham LL, Ashley DL. Using biologic markers in blood to assess exposure to multiple environmental chemicals for inner-city children 3-6 years of age. Environ Health Perspect. 2006;114:453-459.

45. Valcke M, Samuel O, Bouchard M, Dumas P, Belleville D, Tremblay C. Biological monitoring of exposure to organophosphate pesticides in children living in peri-urban areas of the Province of Quebec, Canada. Int Arch Occup Environ Health. 2006;79:568-577.

46. Lewis RG, Fortmann RC, Camann DE. Evaluation of methods for monitoring the potential exposure of small children to pesticides in the residential environment. Arch Environ Contam Toxicol. 1994 Jan;26(1): 37-46.

47. Colt JS, Severson RK, Lubin J, Rothman N, Camann D, Davis S, Cerhan JR, Cozen W, Hartge P. Organochlorines in carpet dust and non-Hodgkin lymphoma. Epidemiology. 2005 Jul;16(4):516-525.

48. California EPA Department of Pesticide Regulation, 1995 Summary of pesticide use report data. Available from: *http://www.cdpr.ca.gov/* [cited 2006 May 23].

49. Gunier RB, Harnly ME, Reynolds P, Hertz A, Von Behren J. Agricultural pesticide use in California: pesticide prioritization, use densities, and population distributions for a childhood cancer study. Environ Health Perspect. 2001 Oct;109(10):1071-1078.

50. Clary T, Ritz B. Pancreatic cancer mortality and organochlorine pesticide exposure in California, 1989-1996. Am J Ind Med. 2003 Mar;43(3):306-313.

51. Teske ME, Bird SL, Esterly DM, Curbishley TB, Ray SL, Perry SG. AgDRIFT: a model for estimating near-field spray drift from aerial applications. Environ Toxicol Chem. 2002 Mar;21(3):659-671.

52. Ward MH, Nuckols JR, Weigel SJ, Maxwell SK, Cantor KP, Miller RS. Identifying populations potentially exposed to agricultural pesticides using remote sensing and a Geographic Information System. Environ Health Perspect. 2000 Jan;108(1):5-12.

53. Rull RP, Ritz B. Historical pesticide exposure in California using pesticide use reports and land-use surveys: an assessment of misclassification error and bias. Environ Health Perspect. 2003 Oct;111(13):1582-1589.

54. Nuckols JR. Agricultural chemical exposures and childhood cancer. Final progress report to the National Cancer Institute. 1 RO3 CA83071-01. 2002. Available from: *http://ehasl.cvmbs.colostate.edu/projects/RO3 Final Progress report.pdf* [cited 2006 May 23].

55. Brody JG, Vorhees DJ, Melly SJ, Swedis SR, Drivas PJ, Rudel RA. Using GIS and historical records to reconstruct residential exposure to large-scale pesticide application. J Expo Anal Environ Epidemiol. 2002 Jan-Feb;12(1):64-80.

56. Simcox NJ, Fenske RA, Wolz SA, Lee IC, Kalman DA. Pesticides in household dust and soil: exposure pathways for children of agricultural families. Environ Health Perspect. 1995 Dec;103(12):1126-1134.

57. Lu C, Fenske RA, Simcox NJ, Kalman D. Pesticide exposure of children in an agricultural community: evidence of household proximity to farmland and take home exposure pathways. Environ Res. 2000 Nov; 84(3):290-302.

58. Loewenherz C, Fenske RA, Simcox NJ, Bellamy G, Kalman D. Biological monitoring of organophosphorus pesticide exposure among children of agricultural workers in central Washington State. Environ Health Perspect. 1997 Dec;105(12):1344-1353.

59. Greenlee RT, Hill-Harmon MB, Murray T, Thun M. Cancer statistics, 2001. CA Cancer J Clin. 2001 Jan-Feb;51(1):15-36.

60. Hsing AW, Tsao L, Devesa SS. International trends and patterns of prostate cancer incidence and mortality. Int J Cancer. 2000 Jan 1;85(1):60-67.

61. Siemiatycki J. Risk factors for cancer in the workplace. Boca Raton (FL): CRC Press; 1991. pp. 276-279.

62. Parent ME, Siemiatycki J. Occupation and prostate cancer. Epidemiol Rev. 2001;23(1):138-143.

63. Morrison H, Savitz D, Semenciw R, Hulka B, Mao Y, Morison D, Wigle D. Farming and prostate cancer mortality. Am J Epidemiol. 1993 Feb 1;137(3): 270-280.

64. Aronson KJ, Siemiatycki J, Dewar R, Gerin M. Occupational risk factors for prostate cancer: results from a case-control study in Montreal, Quebec, Canada. Am J Epidemiol. 1996 Feb 15;143(4):363-373.

65. Blair A, Zahm SH. Agricultural exposures and cancer. Environ Health Perspect. 1995 Nov;103 Suppl 8:205-208.

66. Blair A, Dosemeci M, Heineman EF. Cancer and other causes of death among male and female farmers from twenty-three states. Am J Ind Med. 1993 May; 23(5):729-742.

67. Dosemeci M, Hoover RN, Blair A, Figgs LW, Devesa S, Grauman D, Fraumeni JF Jr. Farming and prostate cancer among African-Americans in the southeastern United States. J Natl Cancer Inst. 1994 Nov 16;86(22):1718-1719.

68. Dich J, Wiklund K. Prostate cancer in pesticide applicators in Swedish agriculture. Prostate. 1998 Feb 1;34(2):100-112.

69. Alavanja MC, Samanic C, Dosemeci M, Lubin J, Tarone R, Lynch CF, Knott C, Thomas K, Hoppin JA, Barker J, Coble J, Sandler DP, Blair A. Use of agricultural pesticides and prostate cancer risk in the Agricultural Health Study cohort. Am J Epidemiol. 2003 May 1;157(9):800-814.

70. Williams MD, Sandler AB. The epidemiology of lung cancer. Cancer Treat Res. 2001;105:31-52.

71. Luchtrath H. The consequences of chronic arsenic poisoning among Moselle wine growers. Pathoanatomical investigations of post-mortem examinations

performed between 1960 and 1977. J Cancer Res Clin Oncol. 1983;105(2):173-182.

72. Mabuchi K, Lilienfeld AM, Snell LM. Cancer and occupational exposure to arsenic: a study of pesticide workers. Prev Med. 1980 Jan;9(1):51-77.

73. Environmental Protection Agency, Office of Pesticide Programs. List of chemicals evaluated for carcinogenic potential. Washington (DC): U S Environmental Protection Agency; 2002. Available from: *http://www.epa.gov/pesticides/carlist/* [cited 2006 May 23].

74. Blair A, Grauman DJ, Lubin JH, Fraumeni JF Jr. Lung cancer and other causes of death among licensed pesticide applicators. J Natl Cancer Inst. 1983 Jul;71(1): 31-37.

75. Pesatori AC, Sontag JM, Lubin JH, Consonni D, Blair A. Cohort mortality and nested case-control study of lung cancer among structural pest control workers in Florida (United States). Cancer Causes Control. 1994 Jul;5(4):310-318.

76. Barthel E. Increased risk of lung cancer in pesticide-exposed male agricultural workers. J Toxicol Environ Health. 1981 Nov-Dec;8(5-6):1027-1040.

77. Becher H, Flesch-Janys D, Kauppinen T, Kogevinas M, Steindorf K, Manz A, Wahrendorf J. Cancer mortality in German male workers exposed to phenoxy herbicides and dioxins. Cancer Causes Control. 1996 May;7(3):312-321.

78. Kogevinas M, Becher H, Benn T, Bertazzi PA, Boffetta P, Bueno-de-Mesquita HB, Coggon D, Colin D, Flesch-Janys D, Fingerhut M, Green L, Kauppinen T, Littorin M, Lynge E, Mathews JD, Neuberger M, Pearce N, Saracci R. Cancer mortality in workers exposed to phenoxy herbicides, chlorophenols, and dioxins. An expanded and updated international cohort study. Am J Epidemiol. 1997 Jun 15;145(12):1061-1075.

79. MacMahon B, Monson RR, Wang HH, Zheng TZ. A second follow-up of mortality in a cohort of pesticide applicators. J Occup Med. 1988 May;30(5):429-432.

80. Wang HH, MacMahon B. Mortality of pesticide applicators. J Occup Med. 1979 Nov;21(11):741-744.

81. Bond GG, Wetterstroem NH, Roush GJ, McLaren EA, Lipps TE, Cook RR. Cause specific mortality among employees engaged in the manufacture, formulation, or packaging of 2,4-dichlorophenoxyacetic acid and related salts. Br J Ind Med. 1988 Feb;45(2):98-105.

82. Coggon D, Pannett B, Winter PD, Acheson ED, Bonsall J. Mortality of workers exposed to 2 methyl-4 chlorophenoxyacetic acid. Scand J Work Environ Health. 1986 Oct;12(5):448-454.

83. Ott MG, Olson RA, Cook RR, Bond GG. Cohort mortality study of chemical workers with potential exposure to the higher chlorinated dioxins. J Occup Med. 1987 May;29(5):422-429.

84. Alavanja MC, Dosemeci M, Samanic C, Lubin J, Lynch CF, Knott C, Barker J, Hoppin JA, Sandler DP, Coble J, Thomas K. Blair A. Pesticides and lung cancer risk in the agricultural health study cohort. Am J Epidemiol. 2004 Nov 1;160(9):876-885.

85. Lee WJ, Blair A, Hoppin JA, Lubin JH, Rusiecki JA, Sandler DP, Dosemeci M, Alavanja MC. Cancer incidence among pesticide applicators exposed to chlorpyrifos in the Agricultural Health Study. J Natl Cancer Inst. 2004 Dec 1;96(23):1781-1789.

86. Jemal A, Tiwari RC, Murray T, Ghafoor A, Samuels A, Ward E, Feuer EJ, Thun MJ; American Cancer Society. Cancer statistics, 2004. CA Cancer J Clin. 2004 Jan-Feb;54(1):8-29.

87. Potter JD. Colorectal cancer: molecules and populations. J Natl Cancer Inst. 1999 Jun 2;91(11):916-932.

88. Swanson GM, Belle SH, Burrows RW Jr. Colon cancer incidence among modelmakers and patternmakers in the automobile manufacturing industry. A continuing dilemma. J Occup Med. 1985 Aug;27(8):567-569.

89. Ehrlich A, Rohl AN, Holstein EC. Asbestos bodies in carcinoma of colon in an insulation worker with asbestosis. JAMA. 1985 Nov 22-29;254(20):2932-2933.

90. Dement J, Pompeii L, Lipkus IM, Samsa GP. Cancer incidence among union carpenters in New Jersey. J Occup Environ Med. 2003 Oct;45(10):1059-1067.

91. Calvert GM, Ward E, Schnorr TM, Fine LJ. Cancer risks among workers exposed to metalworking fluids: a systematic review. Am J Ind Med. 1998 Mar; 33(3):282-292.

92. Tilley BC, Johnson CC, Schultz LR, Buffler PA, Joseph CL. Risk of colorectal cancer among automotive pattern and model makers. J Occup Med. 1990 Jun; 32(6):541-546.

93. Vineis P, Ciccone G, Magnino A. Asbestos exposure, physical activity and colon cancer: a case-control study. Tumori. 1993 Oct 31;79(5):301-303.

94. Alavanja MC, Sandler DP, Lynch CF, Knott C, Lubin JH, Tarone R, Thomas K, Dosemeci M, Barker J, Hoppin JA, Blair A. Cancer incidence in the agricultural health study. Scand J Work Environ Health. 2005;31 Suppl 1:39-45.

95. Forastiere F, Quercia A, Miceli M, Settimi L, Terenzoni B, Rapiti E, Faustini A, Borgia P, Cavariani F, Perucci CA. Cancer among farmers in central Italy. Scand J Work Environ Health. 1993 Dec;19(6):382-389.

96. Zhong Y, Rafnsson V. Cancer incidence among Icelandic pesticide users. Int J Epidemiol. 1996 Dec; 25(6):1117-1124.

97. de Jong G, Swaen GM, Slangen JJ. Mortality of workers exposed to dieldrin and aldrin: a retrospective cohort study. Occup Environ Med. 1997 Oct;54(10): 702-707.

98. Swaen GM, de Jong G, Slangen JJ, van Amelsvoort LG. Cancer mortality in workers exposed to dieldrin and aldrin: an update. Toxicol Ind Health. 2002 Mar;18(2): 63-70.

99. Acquavella JF, Delzell E, Cheng H, Lynch CF, Johnson G. Mortality and cancer incidence among alachlor manufacturing workers 1968-99. Occup Environ Med. 2004 Aug;61(8):680-685.

100. Leet T, Acquavella J, Lynch C, Anne M, Weiss NS, Vaughan T, Checkoway H. Cancer incidence among

alachlor manufacturing workers. Am J Ind Med. 1996 Sep;30(3):300-306.

101. Soliman AS, Smith MA, Cooper SP, Ismail K, Khaled H, Ismail S, McPherson RS, Seifeldin IA, Bondy ML. Serum organochlorine pesticide levels in patients with colorectal cancer in Egypt. Arch Environ Health. 1997 Nov-Dec;52(6):409-415.

102. Lee WJ, Sandler DP, Blair A, Samanic C, Cross AJ, Alavanja MC. Pesticide use and colorectal cancer risk in the agricultural health study. Int J Cancer. 2007 Jul 15;121(2):339-346.

103. Falk RT, Pickle LW, Fontham ET, Correa P, Morse A, Chen V, Fraumeni JJ Jr. Occupation and pancreatic cancer risk in Louisiana. Am J Ind Med. 1990; 18(5):565-576.

104. Forastiere F, Quercia A, Miceli M, Settimi L, Terenzoni B, Rapiti E, Faustini A, Borgia P, Cavariani F, Perucci CA. Cancer among farmers in central Italy. Scand J Work Environ Health. 1993;19:382-389.

105. Milham S Jr. Occupational mortality in Washington State, 1950-1989. DHHS (NIOSH) Publication No. 96-133. Washingon (DC): U.S. Government Printing Office. Available from: *http://www.cdc.gov/niosh/pdfs/96-133.pdf* [cited 2006 May 24].

106. Alavanja MC, Blair A, Masters MN. Cancer mortality in the U.S. flour industry. J Natl Cancer Inst. 1990 May 16;82(10):840-848.

107. Alguacil J, Kauppinen T, Porta M, Partanen T, Malats N, Kogevinas M, Benavides FG, Obiols J, Bernal F, Rifa J, Carrato A. Risk of pancreatic cancer and occupational exposures in Spain. PANKRAS II Study Group. Ann Occup Hyg. 2000 Aug;44(5):391-403.

108. Figa-Talamanca I, Mearelli I, Valente P, Bascherini S. Cancer mortality in a cohort of rural licensed pesticide users in the province of Rome. Int J Epidemiol. 1993 Aug;22(4):579-583.

109. Fryzek JP, Garabrant DH, Harlow SD, Severson RK, Gillespie BW, Schenk M, Schottenfeld D. A case-control study of self-reported exposures to pesticides and pancreas cancer in southeastern Michigan. Int J Cancer. 1997 Jul 3;72(1):62-67.

110. Kauppinen T, Partanen T, Degerth R, Ojajarvi A. Pancreatic cancer and occupational exposures. Epidemiology. 1995 Sep;6(5):498-502.

111. Partanen T, Kauppinen T, Degerth R, Moneta G, Mearelli I, Ojajarvi A, Hernberg S, Koskinen H, Pukkala E. Pancreatic cancer in industrial branches and occupations in Finland. Am J Ind Med. 1994 Jun;25(6): 851-866.

112. Zhong Y, Rafnsson V. Cancer incidence among Icelandic pesticide users. Int J Epidemiol. 1996;25: 1117-1124.

113. Garabrant DH, Held J, Langholz B, Peters JM, Mack TM. DDT and related compounds and risk of pancreatic cancer. J Natl Cancer Inst. 1992 May 20;84(10): 764-771.

114. Beard J, Sladden T, Morgan G, Berry G, Brooks L, McMichael A. Health impacts of pesticide exposure in a cohort of outdoor workers. Environ Health Perspect. 2003 May;111(5):724-730.

115. Clary T, Ritz B. Pancreatic cancer mortality and organochlorine pesticide exposure in California, 1989-1996. Am J Ind Med. 2003 Mar;43(3):306-313.

116. Amoateng-Adjepong Y, Sathiakumar N, Delzell E, Cole P. Mortality among workers at a pesticide manufacturing plant. J Occup Environ Med. 1995 Apr;37(4): 471-478.

117. Coggon D, Pannett B, Winter PD, Acheson ED, Bonsall J. Mortality of workers exposed to 2 methyl-4 chlorophenoxyacetic acid. Scand J Work Environ Health. 1986 Oct;12(5):448-454.

118. Ribbens PH. Mortality study of industrial workers exposed to aldrin, dieldrin and endrin. Int Arch Occup Environ Health. 1985;56(2):75-79.

119. Sathiakumar N, Delzell E, Austin H, Cole P. A follow-up study of agricultural chemical production workers. Am J Ind Med. 1992;21(3):321-330.

120. Wong O, Brocker W, Davis HV, Nagle GS. Mortality of workers potentially exposed to organic and inorganic brominated chemicals, DBCP, TRIS, PBB, and DDT. Br J Ind Med. 1984 Feb;41(1):15-24.

121. Blair A, Zahm SH, Pearce NE, Heineman EF, Fraumeni JF Jr. Clues to cancer etiology from studies of farmers. Scand J Work Environ Health. 1992 Aug; 18(4):209-215.

122. Burmeister LF. Cancer in Iowa farmers: recent results. Am J Ind Med. 1990;18(3):295-301.

123. Franceschi S, Barbone F, Bidoli E, Guarneri S, Serraino D, Talamini R, La Vecchia C. Cancer risk in farmers: results from a multi-site case-control study in north-eastern Italy. Int J Cancer. 1993 Mar 12;53(5): 740-745.

124. Saftlas AF, Blair A, Cantor KP, Hanrahan L, Anderson HA. Cancer and other causes of death among Wisconsin farmers. Am J Ind Med. 1987;11(2):119-129.

125. Keller-Byrne JE, Khuder SA, Schaub EA, McAfee O. A meta-analysis of non-Hodgkin's lymphoma among farmers in the central United States. Am J Ind Med. 1997 Apr;31(4):442-444.

126. Cantor KP, Blair A, Everett G, Gibson R, Burmeister LF, Brown LM, Schuman L, Dick FR. Pesticides and other agricultural risk factors for non-Hodgkin's lymphoma among men in Iowa and Minnesota. Cancer Res. 1992 May 1;52(9):2447-2455.

127. McDuffie HH, Pahwa P, McLaughlin JR, Spinelli JJ, Fincham S, Dosman JA, Robson D, Skinnider LF, Choi NW. Non-Hodgkin's lymphoma and specific pesticide exposures in men: cross-Canada study of pesticides and health. Cancer Epidemiol Biomarkers Prev. 2001 Nov;10(11):1155-1163.

128. Woods JS, Polissar L, Severson RK, Heuser LS, Kulander BG. Soft tissue sarcoma and non-Hodgkin's lymphoma in relation to phenoxyherbicide and chlorinated phenol exposure in western Washington. J Natl Cancer Inst. 1987 May;78(5):899-910.

129. Zahm SH, Weisenburger DD, Babbitt PA, Saal RC, Vaught JB, Cantor KP, Blair A. A case-control

study of non-Hodgkin's lymphoma and the herbicide 2,4-dichlorophenoxyacetic acid (2,4-D) in eastern Nebraska. Epidemiology. 1990 Sep;1(5):349-356.

130. Zheng T, Zahm SH, Cantor KP, Weisenburger DD, Zhang Y, Blair A. Agricultural exposure to carbamate pesticides and risk of non-Hodgkin lymphoma. J Occup Environ Med. 2001 Jul;43(7):641-649.

131. Kogevinas M, Becher H, Benn T, Bertazzi PA, Boffetta P, Bueno-de-Mesquita HB, Coggon D, Colin D, Flesch-Janys D, Fingerhut M, Green L, Kauppinen T, Littorin M, Lynge E, Mathews JD, Neuberger M, Pearce N, Saracci R. Cancer mortality in workers exposed to phenoxy herbicides, chlorophenols, and dioxins. An expanded and updated international cohort study. Am J Epidemiol. 1997 Jun 15;145(12):1061-1075.

132. Groves FD, Linet MS, Travis LB, Devesa SS. Cancer surveillance series: non-Hodgkin's lymphoma incidence by histologic subtype in the United States from 1978 through 1995. J Natl Cancer Inst. 2000 Aug 2;92(15):1240-1251.

133. Nanni O, Amadori D, Lugaresi C, Falcini F, Scarpi E, Saragoni A, Buiatti E. Chronic lymphocytic leukaemias and non-Hodgkin's lymphomas by histological type in farming-animal breeding workers: a population case-control study based on a priori exposure matrices. Occup Environ Med. 1996 Oct;53(10):652-657.

134. Cavel J, Hemon D, Mandereau L Delemotte B, Severin F, Flandrin G. Farming, pesticide use and hairy-cell leukemia. Scad. J. Work Environ Health 1996;22:285-293.

135. Hardell L, Eriksson M, Nordstrom M. Exposure to pesticides as risk factor for non-Hodgkin's lymphoma and hairy cell leukemia: pooled analysis of two Swedish case-control studies. Leuk Lymphoma. 2002 May;43(5):1043-1049.

136. Boros LG, Williams RD. Isofenphos induced metabolic changes in K562 myeloid blast cells. Leuk Res. 2001 Oct;25(10):883-890.

137. Boros LG, Torday JS, Lim S, Bassilian S, Cascante M, Lee WN. Transforming growth factor beta2 promotes glucose carbon incorporation into nucleic acid ribose through the nonoxidative pentose cycle in lung epithelial carcinoma cells. Cancer Res. 2000 Mar 1; 60(5):1183-1185.

138. Stephenson J, Czepulkowski B, Hirst W, Mufti GJ. Deletion of the acetylcholinesterase locus at 7q22 associated with myelodysplastic syndromes (MDS) and acute myeloid leukaemia (AML). Leuk Res. 1996 Mar; 20(3):235-241.

139. Khuder SA, Mutgi AB. Meta-analyses of multiple myeloma and farming. Am J Ind Med. 1997 Nov;32(5):510-516.

140. Cerhan JR, Cantor KP, Williamson K, Lynch CF, Torner JC, Burmeister LF. Cancer mortality among Iowa farmers: recent results, time trends, and lifestyle factors (United States). Cancer Causes Control. 1998 May;9(3):311-319.

141. Lee E, Burnett CA, Lalich N, Cameron LL, Sestito JP. Proportionate mortality of crop and livestock farmers in the United States, 1984-1993. Am J Ind Med. 2002 Nov;42(5):410-420.

142. Eriksson M, Hardell L, Berg NO, Moller T, Axelson O. Soft-tissue sarcomas and exposure to chemical substances: a case-referent study. Br J Ind Med. 1981 Feb;38(1):27-33.

143. Hardell L, Sandstrom A. Case-control study: soft-tissue sarcomas and exposure to phenoxyacetic acids or chlorophenols. Br J Cancer. 1979 Jun;39(6):711-717.

144. Hardell L, Eriksson M. The association between soft tissue sarcomas and exposure to phenoxyacetic acids. A new case-referent study. Cancer. 1988 Aug 1; 62(3):652-656.

145. Vineis P, Terracini B, Ciccone G, Cignetti A, Colombo E, Donna A, Maffi L, Pisa R, Ricci P, Zanini E, Comba P. Phenoxy herbicides and soft-tissue sarcomas in female rice weeders. A population-based case-referent study. Scand J Work Environ Health. 1987 Feb;13(1):9-17.

146. Wingren G, Fredrikson M, Brage HN, Nordenskjold B, Axelson O. Soft tissue sarcoma and occupational exposures. Cancer. 1990 Aug 15;66(4):806-811.

147. Hoar SK, Blair A, Holmes FF, Boysen CD, Robel RJ, Hoover R, Fraumeni JF Jr. Agricultural herbicide use and risk of lymphoma and soft-tissue sarcoma. JAMA. 1986 Sep 5;256(9):1141-1147.

148. Smith AH, Pearce NE, Fisher DO, Giles HJ, Teague CA, Howard JK. Soft tissue sarcoma and exposure to phenoxyherbicides and chlorophenols in New Zealand. J Natl Cancer Inst. 1984 Nov;73(5):1111-1117.

149. Woods JS, Polissar L, Severson RK, Heuser LS, Kulander BG. Soft tissue sarcoma and non-Hodgkin's lymphoma in relation to phenoxyherbicide and chlorinated phenol exposure in western Washington. J Natl Cancer Inst. 1987;78:899-910.

150. Hoppin JA, Tolbert PE, Flanders WD, Zhang RH, Daniels DS, Ragsdale BD, Brann EA. Occupational risk factors for sarcoma subtypes. Epidemiology. 1999 May;10(3):300-306.

151. Dewailly E, Dodin S, Verreault R, Ayotte P, Sauve L, Morin J, Brisson J. High organochlorine body burden in women with estrogen receptor-positive breast cancer. J Natl Cancer Inst. 1994 Feb 2;86(3):232-234.

152. Falck F Jr, Ricci A Jr, Wolff MS, Godbold J, Deckers P. Pesticides and polychlorinated biphenyl residues in human breast lipids and their relation to breast cancer. Arch Environ Health. 1992 Mar-Apr;47(2):143-146.

153. Wolff MS, Toniolo PG, Lee EW, Rivera M, Dubin N. Blood levels of organochlorine residues and risk of breast cancer. J Natl Cancer Inst. 1993 Apr 21;85(8):648-652.

154. Krieger N, Wolff MS, Hiatt RA, Rivera M, Vogelman J, Orentreich N. Breast cancer and serum organochlorines: a prospective study among white, black, and Asian women. J Natl Cancer Inst. 1994 Apr 20;86(8):589-599.

155. Calle EE, Frumkin H, Henley SJ, Savitz DA, Thun MJ. Organochlorines and breast cancer risk. CA Cancer J Clin. 2002 Sep-Oct;52(5):301-309.

156. Collabor Group on Hormonal Factors in Breast Cancer. Breast cancer and hormone replacement therapy: collaborative reanalysis of data from 51 epidemiological studies of 52,705 women with breast cancer and 108,411 women without breast cancer. Lancet 1997 Oct 11;350(9084):1047-1059.

157. Gammon MD, Wolff MS, Neugut AI, Eng SM, Teitelbaum SL, Britton JA, Terry MB, Levin B, Stellman SD, Kabat GC, Hatch M, Senie R, Berkowitz G, Bradlow HL, Garbowski G, Maffeo C, Montalvan P, Kemeny M, Citron M, Schnabel F, Schuss A, Hajdu S, Vinceguerra V, Niguidula N, Ireland K, Santella RM. Environmental toxins and breast cancer on Long Island. II. Organochlorine compound levels in blood. Cancer Epidemiol Biomarkers Prev. 2002 Aug;11(8):686-697.

158. Hunter DJ, Hankinson SE, Laden F, Colditz GA, Manson JE, Willett WC, Speizer FE, Wolff MS. Plasma organochlorine levels and the risk of breast cancer. N Engl J Med. 1997 Oct 30;337(18):1253-1258.

159. Romieu I, Hernandez-Avila M, Lazcano-Ponce E, Weber JP, Dewailly E. Breast cancer, lactation history, and serum organochlorines. Am J Epidemiol. 2000 Aug 15;152(4):363-370.

160. Zheng T, Holford TR, Mayne ST, Ward B, Carter D, Owens PH, Dubrow R, Zahm SH, Boyle P, Archibeque S, Tessari J. DDE and DDT in breast adipose tissue and risk of female breast cancer. Am J Epidemiol. 1999 Sep 1;150(5):453-458.

161. Alavanja MC, Sandler DP, Lynch CF, Knott C, Lubin JH, Tarone R, Thomas K, Dosemeci M, Barker J, Hoppin JA, Blair A. Cancer incidence in the agricultural health study. Scand J Work Environ Health. 2005;31: 39-45.

162. Khuder SA, Mutgi AB, Schaub EA, Tano BD. Meta-analysis of Hodgkin's disease among farmers. Scand J Work Environ Health. 1999 Oct;25(5):436-441.

163. Sharpe RM. Reproductive biology. Another DDT connection. Nature. 1995 Jun 15;375(6532): 538-539.

164. Coleman S, Linderman R, Hodgson E, Rose RL. Comparative metabolism of chloroacetamide herbicides and selected metabolites in human and rat liver microsomes. Environ Health Perspect. 2000 Dec;108(12): 1151-1157.

165. Hu J, Mao Y, White K. Renal cell carcinoma and occupational exposure to chemicals in Canada. Occup Med (Lond). 2002 May;52(3):157-164.

166. Yeni-Komshian H, Holly EA. Childhood brain tumours and exposure to animals and farm life: a review. Paediatr Perinat Epidemiol. 2000 Jul;14(3):248-256.

167. Bohnen NI, Kurland LT. Brain tumor and exposure to pesticides in humans: a review of the epidemiologic data. J Neurol Sci. 1995 Oct;132(2): 110-121.

168. Daniels JL, Olshan AF, Savitz DA. Pesticides and childhood cancers. Environ Health Perspect. 1997 Oct;105(10):1068-1077.

169. Zahm SH, Ward MH. Pesticides and childhood cancer. Environ Health Perspect. 1998 Jun;106 Suppl 3:893-908.

170. Buffler PA, Kwan ML, Reynolds P, Urayama KY. Environmental and genetic risk factors for childhood leukemia: appraising the evidence. Cancer Invest. 2005;23(1):60-75.

171. Efird JT, Holly EA, Preston-Martin S, Mueller BA, Lubin F, Filippini G, Peris-Bonet R, McCredie M, Cordier S, Arslan A, Bracci PM. Farm-related exposures and childhood brain tumours in seven countries: results from the SEARCH International Brain Tumour Study. Paediatr Perinat Epidemiol. 2003 Apr;17(2): 201-211.

172. Kristensen P, Andersen A, Irgens LM, Bye AS, Sundheim L. Cancer in offspring of parents engaged in agricultural activities in Norway: incidence and risk factors in the farm environment. Int J Cancer. 1996 Jan 3;65(1):39-50.

173. Meinert R, Kaatsch P, Kaletsch U, Krummenauer F, Miesner A, Michaelis J. Childhood leukaemia and exposure to pesticides: results of a case-control study in northern Germany. Eur J Cancer. 1996 Oct;32A(11): 1943-1948.

174. Meinert R, Schuz J, Kaletsch U, Kaatsch P, Michaelis J. Leukemia and non-Hodgkin's lymphoma in childhood and exposure to pesticides: results of a register-based case-control study in Germany. Am J Epidemiol. 2000 Apr 1;151(7):639-646.

175. Olshan AF, Daniels JL. Invited commentary: pesticides and childhood cancer. Am J Epidemiol 2000; 151:647-649.

176. Flower KB, Hoppin JA, Lynch CF, Blair A, Knott C, Shore DL, Sandler DP. Cancer risk and parental pesticide application in children of Agricultural Health Study participants. Environ Health Perspect. 2004 Apr;112(5):631-635.

177. Bonner MR, Alavanja MC. The Agricultural Health Study biomarker workshop on cancer etiology. Introduction: overview of study design, results, and goals of workshop. J Biochem Mol Toxicol. 2005; 19(3):169-171.

178. Guengerich FP. Generation of reactive intermediates. J Biochem Mol Toxicol. 2005;19(3):173-174.

179. Selgrade MK. Biomarkers of effects: the immune system. J Biochem Mol Toxicol. 2005;19(3): 177-179.

180. Burns-Naas LA, Meade BJ, Munson AE. Toxic responses of the immune system. In: Klaassen, CD. Casarett & Doull's Toxicology: The basic science of poisons. 6th ed. New York (NY): McGraw-Hill; 2001. pp. 419-470.

181. Flipo D, Bernier J, Girard D, Krzystyniak K, Fournier M. Combined effects of selected insecticides on humoral immune response in mice. Int J Immunopharmacol. 1992 Jul;14(5):747-752.

182. Dewailly E, Ayotte P, Bruneau S, Gingras S, Belles-Isles M, Roy R. Susceptibility to infections and

immune status in Inuit infants exposed to organochlorines. Environ Health Perspect. 2000 Mar;108(3):205-211.

183. Arndt V, Vine MF, Weigle K. Environmental chemical exposures and risk of herpes zoster. Environ Health Perspect. 1999 Oct;107(10):835-841.

184. Luebke RW, Parks C, Luster MI. Suppression of immune function and susceptibility to infections in humans: association of immune function with clinical disease. J Immunotoxicol. 2004;1(1):15-24.

185. Young LS, Rickinson AB. Epstein-Barr virus: 40 years on. Nat Rev Cancer. 2004 Oct;4(10):757-768.

186. Moore LE, Pfeiffer R, Warner M, Clark M, Skibola C, Steinmous C, Alguacil J, Rothman N, Smith MT, Smith AH. Identification of biomarkers of arsenic exposure and metabolism in urine using SELDI technology. J Biochem Mol Toxicol. 2005;19(3):176.

187. Furlong CE, Cole TB, Walter BJ, Shih DM, Tward A, Lusis AJ, Timchalk C, Richter RJ, Costa LG.

Paraoxonase 1 (PON1) status and risk of insecticide exposure. J Biochem Mol Toxicol. 2005;19(3):182-183.

188. Nebert DW. Role of host susceptibility to toxicity and cancer caused by pesticides: cytochromes P450. J Biochem Mol Toxicol. 2005;19(3):184-186.

189. Eaton DL. Glutathione S-transferases as putative host susceptibility genes for cancer risk in agricultural workers. J Biochem Mol Toxicol. 2005;19(3):187-189.

190. Kunkel TA. DNA replication and repair reactions relevant to the AHS. J Biochem Mol Toxicol. 2005;19(3):190-191.

191. Garry VF. Biomarkers of thyroid function, genotoxicity and agricultural fungicide use. J Biochem Mol Toxicol. 2005;19(3):175.

192. Alavanja MC, Sandler DP, McMaster SB, Zahm SH, McDonnell CJ, Lynch CF, Pennybacker M, Rothman N, Dosemeci M, Bond AE, Blair A. The Agricultural Health Study. Environ Health Perspect. 1996 Apr;104(4):362-369.

doi:10.1300/J096v12n01_05

Surveillance for Pesticide-Related Disease

Ana Maria Osorio, MD, MPH

SUMMARY. Public health surveillance for acute pesticide intoxications is discussed. Explanation of the goals, components and functions of population-based surveillance is provided with reference to key informational sources. Both a state-based pesticide intoxication program and a nearly nationwide poison control center data base program are used to illustrate the potential uses inherent in these types of system. There is additional discussion on the investigation of disease clusters, the use of complementary exposure monitoring data and confidentiality issues. doi:10.1300/J096v12n01_06 *[Article copies available for a fee from The Haworth Document Delivery Service: 1-800-HAWORTH. E-mail address: <docdelivery@haworthpress.com> Website: <http://www.HaworthPress.com>.]*

KEYWORDS. Pesticides, pesticide-related disease, disease clusters, population-based surveillance, poison control, exposure monitoring

INTRODUCTION

Public health surveillance is defined as the ongoing systematic collection, analysis, interpretation and dissemination of information about a health event with the ultimate intent of reducing morbidity and mortality. Specific activities associated with surveillance systems include case outbreak detection, assessment of disease impact and natural history, generation of hypotheses for further research, and evaluation of preventive and control measures.[1] In recent years, syndromic surveillance systems have been implemented for the early detection of illnesses that may result from terrorism-related epidemics. These systems may include monitoring of disease-related events such as clusters of symptoms or medication purchases.[2] Certain components of a surveillance system may be applied to a workplace or community-based medical monitoring program. This type of monitoring program consists of periodic assessment of a specific population which may include a medical interview or questionnaire, physical examination, and laboratory tests that are usually coupled with environmental sampling.

For the purposes of this discussion, surveillance will refer to the public health surveillance model that tracks acute pesticide intoxication. Other research methods are available for chronic and long-term types of pesticide intoxication, but public health surveillance is better suited for the acute type of illness. This review will not

Ana Maria Osorio is Regional Medical Officer for the Pacific Region of the U.S. Food and Drug Administration.

Address correspondence to: Ana Maria Osorio, MD, MPH, U.S. Food and Drug Administration–Pacific Regional Office, 1301 Clay Street, Suite 1180N, Oakland, CA 94612 (E-mail: anamaria.osorio@fda.hhs.gov).

The opinions and contents of this article are those of Dr. Osorio and do not necessarily reflect the views of the U.S. Food and Drug Administration.

[Haworth co-indexing entry note]: "Surveillance for Pesticide-Related Disease." Osorio, Ana Maria. Co-published simultaneously in *Journal of Agromedicine* (The Haworth Medical Press, an imprint of The Haworth Press, Inc.) Vol. 12, No. 1, 2007, pp. 57-66; and: *Proceedings from the Medical Workshop on Pesticide-Related Illnesses from the International Conference on Pesticide Exposure and Health* (ed: Ana Maria Osorio, and Lynn R. Goldman) The Haworth Medical Press, an imprint of The Haworth Press, Inc., 2007, pp. 57-66. Single or multiple copies of this article are available for a fee from The Haworth Document Delivery Service [1-800-HAWORTH, 9:00 a.m. - 5:00 p.m. (EST). E-mail address: docdelivery@haworthpress.com].

include syndromic surveillance, nor be an exhaustive listing of all existing surveillance systems. The intent is to provide several public health surveillance models and resource materials for tracking pesticide intoxications that could be adapted for specific national or international regions: (1) the California-based acute pesticide intoxication program; (2) National Institute for Occupational Safety and Health and Environmental Protection Agency-coordinated surveillance system standards; and (3) a nearly-nationwide poison control center data collection system. In addition, information is provided on investigation of disease clusters identified via surveillance systems, exposure monitoring as a complement to the traditional disease tracking efforts, and confidentiality issues.

COMPONENTS OF A SURVEILLANCE SYSTEM

Acute pesticide intoxication is a health event with significant public health importance that merits consideration as a surveillance system target condition. This type of intoxication has a relatively high incidence rate with potentially severe health effects (especially in high-risk populations), is a preventable condition with effective interventions at the individual and community level, and attracts much public interest with respect to possible health effects. The objectives for this type of surveillance system would include detecting disease outbreaks, monitoring disease trends, developing research studies and evaluating any intervention or control efforts.

For the surveillance system to be effective, certain aspects need to be established: (1) A case definition with person, time and place characteristics that allow a reported case to be confirmed as an acute pesticide intoxication; (2) A standardized information flow (Figure 1); (3) A target population defined by geography, gender, age, workplace, and possibly other descriptive factors; (4) The data collection period; (5) Standardized data collection with list of variables; (6) Data sources identified for case reporting; (7) The standardized data management protocol with attention to confidentiality, handling, storage and security; (8) The analyti-

FIGURE 1. Data flow diagram for a typical acute pesticide intoxication surveillance system. Important to specify target population, geographic area, and time period. Overall system should be evaluated every two years.

cal methods protocol; (9) The reporting protocol including periodic publication of findings; (10) The dissemination plan for findings to reach key stakeholders; and (11) An evaluation plan should occur periodically.[1] The evaluation plan is a key component of the system, and usually addresses the following characteristics: simplicity and flexibility of operations; acceptability by participants; how representative the results are in relation to the target population; timeliness of reporting; stability of the overall system; and the sensitivity and predictive value positive of the cases identified (Table 1).

Acute Pesticide Intoxication Surveillance–California

For an example of a longstanding acute pesticide intoxication surveillance system, consider the California Pesticide Illness Surveillance Program (PISP) within the California Environmental Protection Agency. In California, physician reporting of pesticide illnesses has been required since 1971. A pesticide is any substance that controls pests such as insects, fungi, weeds, rodents, nematodes, algae, viruses, bacteria or adjuvants (chemicals added to enhance efficacy of pesticides). Physicians

TABLE 1. For a disease surveillance system, the sensitivity of case reporting measures how well the number of identified cases reflects the actual number of cases in the target population. The predictive value of a positive case (predictive value positive or PVP) identifies the degree to which a surveillance case will represent a true case (adapted from Ref. 1). $\text{Sensitivity} = \dfrac{A}{(A + C)}$ $\quad \text{PVP} = \dfrac{A}{(A + B)}$

	True Case	True Non-Case	Total
Surveillance Case	A	B	A + B
Surveillance Non-Case	C	D	C + D
Total	A + C	B + D	A + B + C + D

must report any suspected case of pesticide-related illness to their local health department within 24 hours of examining the patient. The local health department then contacts the county agricultural commissioner and completes a Pesticide Illness Report which is reviewed by PISP staff. To identify additional pesticide intoxications, a review of illness reports submitted to the California workers' compensation system is conducted. During certain years, research funds have been provided to incorporate poison control center reports. An investigation is conducted by the county agricultural commissioner staff when a report mentions a specific pesticide as a cause of injury.[3]

Data from these reports and investigations are reviewed by PISP staff. Each reported case is categorized based on the strength of the relationship between the exposure and the resulting illness:[3]

Strength of Relationship for Pesticide Intoxication Case Report	
DEFINITE	High degree of correlation with medical status (e.g., cholinesterase inhibition, positive allergy test, clinician observed physical signs) and exposure (e.g., environmental or biological samples, and exposure history).
PROBABLE	Relatively high degree of correlation with either medical or physical evidence being inconclusive or unavailable.
POSSIBLE	Some degree of correlation with both medical and physical evidence inconclusive or unavailable.

Not related	Inadequate data or cases determined to be unlikely, unrelated or asymptomatic

In Table 2, the five-year cumulative results for PISP are shown. Of the 4,455 cases meeting the case definition with a definite, probable or possible relationship to pesticide exposure, nearly 60% were men. The majority of the cases involved systemic disease and approximately 75% of all cases occurred in the workplace. Of the 13 fatalities, 70% were suicides. The two most frequent activities at the time of exposure were pesticide application and field work (20% and 18%, respectively). Looking more closely at the age trends over this five-year period, the overall peak in age is between 20 and 50 years old (Figure 2). However, the age curve tends to peak a decade earlier when one looks only at the agricultural cases reported.

For the five-year period of 2000-2004, 62% of the 2,752 cases identified by the PISP presented with the more severe systemic or respiratory symptoms (Table 2). In looking at specific pesticide subgroups, the degree of severe clinical manifestations becomes more pronounced: 93% of organophosphate cases, 82% of carbamate cases, and 77% of pyrethrin or pyrethroid cases exhibited systemic or respiratory symptoms. Another way to look at the effect of cholinesterase inhibiting pesticides such as organophosphates and carbamates is shown in Figure 3. There is a nearly twofold increase in the extent of systemic disease when comparing all the cases versus those cases associated with cholinesterase inhibiters: 48% for all pesticide exposures, and 83% for cholinesterase-inhibiting exposures.

Because of the extensive and long duration of this surveillance system, the PISP database

TABLE 2. Selected data from the California Pesticide Illness Surveillance Program for the years 2000 through 2004 (adapted from Ref. 3).

Year	2000	2001	2002	2003	2004	2000-2004 Period (% of total cases)
Total reports received	1144	979	1859	1232	1238	6452
Pesticide related cases[1] Number of men	893 494	616 375	1316 665	802 450	828 560	4455 2544 (57.1)
Category of cases[1] Definite/Probable Possible	 637 256	 430 186	 1025 291	 614 188	 552 276	 3258 (73.1) 1197 (26.9)
Hospitalizations	36	29	25	9	14	113 (2.5)
Fatalities Suicides Unintentional	 5 0	 0 0	 0 3	 4 1	 0 0	 9 (0.2) 4 (0.1)
Health effects of cases[3] Systemic Respiratory Dermal/ocular	 503 103 288	 284 103 229	 592 221 503	 302 141 359	 436 68 323	 2117 (47.5) 636 (14.3) 1702 (38.2)
Circumstances of exposure[2] Agricultural exposure Occupational setting	 417 656	 192 408	 702 1025	 405 553	 390 757	 2106 (47.3) 3399 (76.3)
Selected activities among occupationally exposed Pesticide application Field work Mixing/loading pesticides Packaging/processing Transport/storage/disposal Emergency response Manufacturing/formulation	 175 161 54 42 28 10 2	 132 57 52 11 23 10 3	 168 240 69 137 19 5 5	 196 81 65 18 14 12 3	 196 269 71 31 17 2 5	 867 (19.5) 808 (18.1) 311 (7.0) 239 (5.4) 101 (2.3) 39 (0.9) 18 (0.4)

[1]Pesticide related indicates that the relationship between pesticide exposure and resulting symptoms was one of the following:
   Definite–High degree of correlation with both medical evidence (e.g., cholinesterase inhibition, positive allergy test, signs observed by clinician) and physical evidence of exposure (e.g., environmental or biological samples and exposure history).
   Probable–Relatively high degree of correlation with either medical or physical evidence being inconclusive or unavailable.
   Possible–Some degree of correlation with both medical and physical evidence inconclusive or unavailable.
[2]Categories not mutually exclusive.
[3]One case had unknown health effects.

has been used for health policy development, rule making, research, risk assessment, intervention programs, and evaluation efforts at the state and national level. Like all surveillance systems, there are potential situations which may lead to underreporting of cases: lack of physician recognition of acute intoxication, subtle and early manifestations of intoxications may cause an individual to not seek medical attention; residents or workers may have no or inadequate medical insurance and only seek medical care when symptoms become severe; some residents or workers lack residency documents and do not feel empowered to complain about health problems or seek medical care; some individuals are bi-national and receive medical attention in their native country (e.g., Mexico); and, the migratory and seasonal mobility of some individuals do not allow follow-up or continuity of medical care to allow appropriate diagnosis. With respect to the surveillance sys-

tem, there are some intrinsic aspects that could pose a problem: (1) The strictness of the case definition that may lead to falsely rejecting some cases that are true intoxications; (2) Language and cultural barriers at the employer, medical community and surveillance staff level may pose a problem in fully investigating outbreaks, (especially in high risk groups); and (3) Some of the lower priority pesticide exposure investigations take a longer time to be initiated, which may compromise data collection.

A recent article evaluated the strengths and limitations of this surveillance system by comparing identified cases to those found in California-based records such as death certificates, poison control center logs and hospital discharge data.[4] For the study period 1994-1996, few domestic non-agricultural pesticide intoxications, and only half of agricultural cases, were reported to the surveillance system. The number of individuals affected by an exposure

FIGURE 2. For the years 2000 through 2004, the cumulative number of acute pesticide intoxications by age range and agricultural origin (Ag) are shown. Source: California Pesticide Illness Program (adapted from Ref. 3).

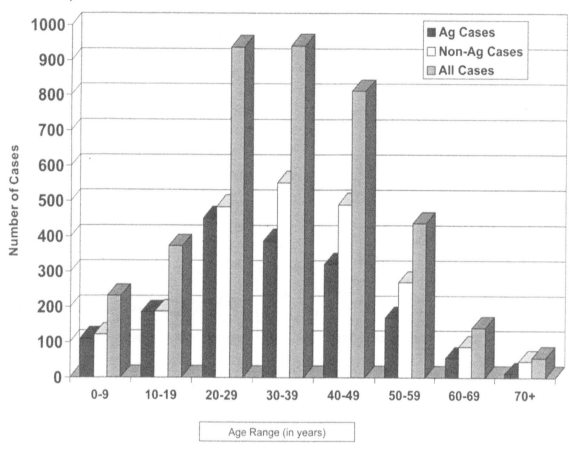

TABLE 3. Types of pesticides and symptoms associated with cases identified from California Pesticide Illness Surveillance Program for the years 2000 through 2004 (adapted from Ref. 3).

Pesticide-Related Cases[1]	Year					2000-2004 Time Period
	2000	2001	2002	2003	2004	
Total cases identified	893	616	1316	802	828	4455
Cases with respiratory or systemic symptoms (%)	605 (67.7)	387 (62.8)	813 (61.8)	443 (55.2)	504 (60.9)	2752 (61.8)
Cases associated with organophosphates (OP)	156	56	84	15	161	472
Of these, number with respiratory or systemic symptoms (% of OP cases)	146 (93.6)	46 (82.1)	76 (90.5)	13 (86.7)	158 (98.1)	439 (93.0)
Cases associated with carbamates	8	8	6	8	3	33
Number with respiratory or systemic symptoms (% of carbamate cases)	6 (75.0)	5 (62.5)	5 (83.3)	8 (100)	3 (100)	27 (81.8)
Cases associated with pyrethrins or pyrethroids (PYR)	64	35	40	26	23	188
Of these, number with respiratory or systemic symptoms (% of PYR cases)	57 (89.1)	26 (74.3)	29 (72.5)	14 (53.8)	19 (82.6)	145 (77.1)

[1]Pesticide related indicates that the relationship between pesticide exposure and resulting symptoms was one of the following:
Definite–High degree of correlation with both medical evidence (e.g., cholinesterase inhibition, positive allergy test, signs observed by clinician) and physical evidence of exposure (e.g., environmental or biological samples and exposure history).
Probable–Relatively high degree of correlation with either medical or physical evidence being inconclusive or unavailable.
Possible–Some degree of correlation with both medical and physical evidence inconclusive or unavailable.

FIGURE 3. For the years 2000 through 2004, the annual number of acute pesticide intoxications by type of illness (systemic, respiratory and dermatologic/ocular symptoms) and association with cholinesterase inhibiting pesticides (ChE) are shown. Source: California Pesticide Illness Program (adapted from Ref. 3).

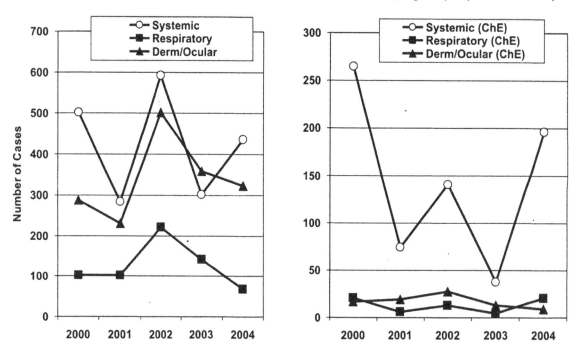

Year Case Reported

was an important factor in whether a case was reported. There was 100% identification of cases in disease clusters involving at least five individuals, and only 16% ascertainment for non-agricultural occupational cases involving four or less exposed individuals. Furthermore, pediatric and intentional exposure cases were under-represented in the surveillance system. The authors conclude that the distribution of pesticide-related health events based on the California surveillance system under-represents those intoxications in the non-occupational and non-agricultural settings.

Other State-Based Surveillance–
United States

In the U.S., there is no nationwide surveillance system for pesticide intoxications. As has been discussed earlier, the PISP is one of the longest running systems in the country. However, there is a second acute pesticide surveillance system in California (of shorter duration that is managed by the state health department) and 11 other states: Arizona, Florida, Iowa,

Louisiana, Massachusetts, Michigan, New Mexico, New York, Oregon, Texas, and Washington. The National Institute for Occupational Safety and Health (NIOSH), in collaboration with the U.S. Environmental Protection Agency (EPA), provides funding for eight of these states and coordinates standardization of data collection and reporting among all 13 programs in 12 states. NIOSH publications include a guidance document on developing a state-based pesticide intoxication surveillance program, standardized case definition and severity index documentation and a computer software program for data entry and analysis.[5]

Estimated Number of Pesticide
Intoxications–United States

In an attempt to get a nationwide estimate of pesticide intoxications, poison control center data has been utilized. Since 1983, the Toxic Exposure Surveillance System (TESS) has collected data from poison control centers and is administered by the American Association of Poison Control Centers.[6] In 2002, there were 64 participating poison control centers with 2.38 million human exposure case reports. The participating centers serve a population of 291.6 million: 49 states (all except for North Dakota), District of Columbia and Puerto Rico. Thus, 99.8% of the U.S. population was served by these centers. However, there is considerable variability in the degree of penetrance for each center (penetrance being the number of human poison exposure cases reported per 1,000 invidivuals per year). Although the majority of the calls deal with drugs, there were 96,112 reports associated with pesticides (4.0% of all human exposures). In Table 4, the most frequent pesticide subgroups among the 96,112 annual pesticides cases were insecticides (52%), rodenticides (21%), and repellents (14%). The identified pesticide cases tend to be very young with 53% less than 6 years of age. However, this trend is driven by the disproportionately high rate of rodenticide cases among the very young. Ninety-four percent of the pesticide cases were due to unintentional causes (predominately insecticides) and 21% resulted in treatment at a health care facility. Moderate and major clinical outcomes affected 3% of all pesticide cases with a total of 18 deaths. Within the insecticide category, there is a higher rate of adults (96% were > 19 years) with 37% of the cases associated with pyrethrins/pyrethroids, and 27% with cholinesterase inhibiting insecticides.

INVESTIGATION
OF DISEASE CLUSTERS

An important activity associated with surveillance systems is the rapid identification and investigation of disease outbreaks or clusters. These investigations can be time-consuming and often require a team approach with various health and safety professionals. To prioritize and focus resources on the high risk situations, a protocol is usually developed with action criteria for when a follow-up investigation should occur.[7] The major steps involved in investigating a disease outbreak include: (1) Formulate a tentative medical diagnosis–confirm diagnosis of index cases, review their medical record and biological tests, review decontamination procedure for patient and consult with treating health care providers; (2) Identify unrecognized cases–interview involved individuals, contact health care providers and medical facilities in area of event to detect other cases, and for occupational settings, interview co-workers, management, union representative (if present), and company medical care provider, as well as review any existing monitoring records; (3) Develop working case definition–see description below; (4) Evaluate the exposure information–formulation and physical form of pesticides, timing and duration of exposure, delivery system, personal protective equipment, weather conditions and general circumstances of event. How many persons were involved? Inspect any machines, equipment, containers, labels, and personal protective equipment that are available. Review any past pesticide exposure or biological monitoring results and environmental sampling records; (5) Create a listing of index cases, additional cases and exposed individuals–note pertinent traits for each person such as age, gender, residence, employment, etc., that may indicate common factors; (6) Plot out the cases to determine the outbreak trend–this epidemic curve results form plotting the cases on an x-axis with time units and a y-axis with number of cases; (7) Determine whether a dose-response relationship exists–cases with an increase in degree or duration of pesticide exposure exhibit a more severe clinical presentation, and, finally; (8) Determine the incidence rate for the outbreak − Incidence = [(number of cases/number of persons exposed) × 100]. Compare this rate to that of the general population for that given medical condition. Sometimes, the incidence rate of a group of non-exposed individuals is calculated. In this situation, a statistical comparison between the exposure incidence rate versus the non-exposed one is made to determine if there is a true outbreak with concurrent elevated risk.

Components of a Working Case Definition
(a) Exposure setting which identifies one or more suspected agents or high-risk activity, population group, location and time period
(b) Anticipated physical signs and symptoms
(c) Pertinent confirmatory biological or environmental tests

TABLE 4. For year 2000, selected characteristics are shown for poison control center reports involving pesticides. Source: TESS (adapted from Ref. 7).

Pesticide Category	Number of cases exposed	Age (in years)			Reason[1]			Number treated in HCF[2]	Case Outcome[3]				
		< 6	6-19	> 19	Unint	Int	Other		None	Minor	Moderate	Major	Death
All pesticides	96,112	50,415	8,342	36,465	90,696	2,457	2,699	19,921	21,844	14,563	2,661	274	18
Fungicides	1,317	359	92	1,252	1,252	21	41	322	189	272	52	3	0
Fumigants	680	71	85	668	668	4	7	147	101	133	32	7	1
Herbicides	9,520	2,508	968	9,032	9,032	117	343	2,177	1,919	2,265	313	20	5
Insecticides	50,134	21,016	4,954	48,053	48,053	1,266	1,660	9,442	9,643	9,132	1,884	167	7
Repellents	12,954	8,932	1,575	12,380	12,380	156	391	1,545	3,241	2,236	204	13	0
Rodenticides	20,507	17,529	668	2,176	19,311	893	257	6,288	6,751	425	176	64	5
Subgroups within the insecticide category[4]													
Carbamates	3,968	1,580	366	1,993	3,726	117	113	779	724	623	135	21	0
OP	9,411	2,739	826	5,723	8,877	241	253	2,390	1,933	1,823	495	67	5
OP/Carb	211	58	19	134	200	9	2	42	37	43	9	1	0
Pyreth	18,442	6,476	2,193	10,573	18,032	532	835	3,503	3,259	4,551	824	38	0
Other	18,102	10,163	1,550	6,198	17,218	367	457	2,728	3,690	2,092	421	40	2

1–Reasons reported include Unint (unintentional), Int (intentional), and Other (other reasons).
2–HCF = Health care facility
3–Medical outcome of case classified as None (no symptoms or physical signs), Minor (minimal symptoms/signs with no residual disability), Moderate (symptoms/signs more pronounced, prolonged or systemic with no residual disability and usually with treatment indicated), Major (symptoms/signs life-threatening or result in permanent disability or disfigurement), or Death.
4–Insecticide category subgroups include: OP (organophosphates), Carb (carbamates), OP/Carb (mixed OP and Carb exposures), Pyreth (pyrethrins and pyrethroids), and Other (other types of pesticides).

Once a potential outbreak is proven to be a true hazard, the information needs to be conveyed to the proper parties to ensure that all cases are appropriately cared for and that any mitigation measures are instituted. This information needs to be distributed to all relevant stakeholders such as community residents, workers, management, medical community, and governmental agencies. The outbreak investigation report should clearly state the extent of the disease, the etiological factors and how to prevent or control the present exposure situation, and recommendations for ways to prevent such episodes in the future. While all individuals evaluated should receive their personal test results, one should ensure that all public reports and communications with stakeholders maintain the confidentiality of the individual participants.

EXPOSURE MONITORING

Although the focus has been on disease surveillance systems, there are population-based exposure monitoring programs that complement disease-based efforts. One example is the ongoing biomonitoring survey conducted by the National Center for Environmental Health within Centers for Disease Control and Prevention (CDC).[8] This program conducts exposure assessment by analyzing selected toxicants in biological samples from the general population. The Third National Report on Human Exposure to Environmental Chemicals provides the latest findings of the ongoing assessment of 148 chemicals. The pesticide groups covered by this survey include: 16 organochlorines; organophosphates (6 dialkyl phosphates, and 6 sp carbamates); DEET; ortho-phenylphenol; and 2,5-dichlorphenol.

The survey obtained urine samples from the National Health Nutrition and Examination Survey (NHANES) for the years 2001-2002. This survey is a continuous national sample of the U.S. population. NHANES conducts a standardized health interview, physical examination, and testing of biological fluids. For organophosphates, the assessment included various urinary metabolites. Of interest is the metabolite indicative of chlorpyrifos and chlorpyrifos-methyl: 3,5,6-trichloro-2-pyridinol (TCP). While use is restricted in the home environment, chlorpyrifos and diazinon are still used extensively in agriculture (especially chlorpyrifos).[8] There does not appear to be much difference in concentration by gender, but there is a threefold increase in child urinary levels as compared to adults. This type of pesticide exposure trend at relatively modest levels in the population would only have been identified using the more precise biological exposure assessments seen in the National Exposure Survey. This exposure tool promises to be a more precise assessment of population body burden than the usual measurement of pesticide sales or agricultural pesticide applications available in some states.

CONFIDENTIALITY CONCERNS

While the use of disease surveillance systems provides the opportunity to better understand disease association risks, it raises special concerns about the confidentiality of the identity of individuals in each record system, especially those that involve linkage between a cohort and outcome file. These are important concerns with the enabling legislation for most population-based disease surveillance systems including provisions for maintaining confidentiality. Because of the possible breach of confidentiality, many Scandinavian countries have enacted strict restrictions on the uses of surveillance data that virtually preclude access to many potential investigators. With the increasing sophistication of geographic information system technology, the potential to identify individuals from characteristics associated with rare events at smaller levels of geographic detail has also caused concern. There are confidentiality concerns about mapped data presentation even when individual identities are unknown. Thus, restrictions have been placed on precise geographic information for rare events derived from surveillance systems.

In 1996, the Health Insurance Portability and Accountability Act (HIPAA) was enacted in the U.S. This regulation provides protection for personal health information (individually identifiable health data). Public health surveillance uses this type of information extensively. Detailed discussion of the applicability of HIPAA for public health practice is contained in a guidance document issued by CDC, and informa-

tion on this topic is found on the Department of Health and Human Services Web site (maintained by the Office of Civil Rights).[9,10]

CONCLUSION

Public health surveillance is not new. Some idea of a defined population at risk, and the incidence of disease in that population, dates back to the earliest efforts to control the spread of communicable diseases. Similar tools have been applied to understanding the scope of non-infectious diseases, such as pesticide intoxication. Vital records provide the most basic model for population-based disease surveillance. Similarly, hospital-based or workplace-centered disease monitoring can provide useful information. However, population-based pesticide-related disease surveillance used in conjunction with geographic, occupational, or environmental information on defined cohorts of interest provide a less biased view of the underlying disease experience. Disease surveillance systems are somewhat limited by the breadth and depth of information on diagnostic detail and relevant risk factors. By virtue of their standards for completeness and consistency, however, they offer a valuable means for assessing important population variations in disease that can be used as the basis for designing more in-depth epidemiologic studies of disease etiology or intervention efforts for disease control.

REFERENCES

1. CDC. Updated guidelines for evaluating public health surveillance systems. MMWR 50 (No. RR-13), 2001.

2. CDC. Framework for evaluating public health surveillance systems for early detection of outbreaks. MMWR (No. RR-5), 2004.

3. Department of Pesticide Regulation, California Environmental Protection Agency. Pesticide Illness Surveillance Program [cited 2007 Feb 6]. Available from: www.cdpr.ca.gov/docs/whs/pisp.htm.

4. Mehler LN, Schenker MB, Romano PS and Samuels SJ. California Surveillance for Pesticide-related Illness and Injury: Coverage, Bias and Limitations. J Agromedicine. 2006; 11(2):67-76.

5. National Institute for Occupational Safety and Health, CDC. Pesticide Illness and Injury Surveillance

[cited 2007 Feb 6]. Available from: www.cdc.gov/niosh/topics/pesticides.

6. Watson WA, Litovitz TL, Rodgers GC, Klein-Schwartz W, Youniss J, Rutherford R, Borys D and May ME. 2002 Annual Report of the American Association of Poison Control Centers Toxic Exposure Surveillance System. Am J Emerg Med. 2003;21(5): 353-421.

7. Osorio AM and Reynolds P. Disease Surveillance Systems. In: LaDou J, editor. Current Occupational and Environmental Medicine, 3rd edition. New York: Lange Medical Books/McGraw-Hill; 2006. pp. 757-74.

8. National Center for Environmental Health, CDC. Third National Report on Human Exposure to Environmental Chemicals [cited 2007 Feb 6]. Available from: www.cdc.gov/exposurereport/3rd/.

9. Department of Health and Human Services, Office of Civil Rights [cited 2007 Feb 6]. Available from: www.hhs.gov/ocr/hipaa/.

10. National Centers for Disease Control and Prevention: HIPAA privacy rule and public health 2003;52:1 [cited 2007 Feb 6]. available from: *www.cdc.gov/mmwr/preview/mmwr/m2e411a1.htm.*

doi:10.1300/J096v12n01_06

Managing Pesticide Chronic Health Risks: U.S. Policies

Lynn R. Goldman, MD, MPH

SUMMARY. This paper provides an overview of U.S. government pesticide risk management efforts over time and in recent years, relevant to chronic health risks of pesticides. Pesticides are in widespread usage in the U.S. With hundreds of active ingredients and thousands of products on the market, management of pesticide risks has been a daunting challenge. The first legislation providing federal authority for regulating pesticides was enacted in 1910. With the establishment of the U.S. Environmental Protection Agency in 1970 and amendments to the pesticide law in 1972, the federal government was for the first time given the authority to regulate health and environmental risks of pesticides. However, older pesticide risks were not addressed until legislation was enacted in 1988, requiring "reregistration" and 1996, requiring that pesticide food standards are safe for children. In result, the U.S. has seen an expansion of development of pesticide products that are registered as "reduced risk" or are biologicals. Additionally a large number of older pesticides have been cancelled or reduced from the market and/or from individual food uses. Through biomonitoring data, the U.S. may now be seeing trends in reduction of exposure to certain pesticides, the organophosphate insecticides. However, pesticide sales data through 2001 do not provide evidence for such trends. doi:10.1300/J096v12n01_07 *[Article copies available for a fee from The Haworth Document Delivery Service: 1-800-HAWORTH. E-mail address: <docdelivery@haworthpress. com> Website: <http://www.HaworthPress.com> © 2007 by The Haworth Press, Inc. All rights reserved.]*

KEYWORDS. Pesticides, pesticide risk management, biomonitoring, biologicals

INTRODUCTION

Even in the U.S., identification and control of potential chronic health risks of pesticides is a daunting task. For perspective, in the United States in 1995, there were 876 pesticides on the market. In 2001, there were nearly 5 billion pounds of pesticides used, of which about 1.2 billion pounds were agricultural and household pest control agents.[1] (Most of the remainder were drinking water disinfectants, wood preservatives, and industrial biocides.) Usage pat-

Lynn R. Goldman is Professor, Environmental Health Sciences, Johns Hopkins Bloomberg School of Public Health, Baltimore, MD.

Address correspondence to: Lynn R. Goldman, MD, MPH, Professor, Environmental Health Sciences, Johns Hopkins Bloomberg School of Public Health, 615 North Wolfe Street, Room E6636, Baltimore, MD 20815 (E-mail: lgoldman@jhsph.edu).

Disclaimer: Although the author served as Assistant Administrator for the US Environmental Protection Agency Office of Prevention, Pesticides and Toxic Substances, this article reflects her own views and not the views of the US government.

[Haworth co-indexing entry note]: "Managing Pesticide Chronic Health Risks: U.S. Policies." Goldman, Lynn R. Co-published simultaneously in *Journal of Agromedicine* (The Haworth Medical Press, an imprint of The Haworth Press, Inc.) Vol. 12, No. 1, 2007, pp. 67-75; and: *Proceedings from the Medical Workshop on Pesticide-Related Illnesses from the International Conference on Pesticide Exposure and Health* (ed: Ana Maria Osorio, and Lynn R. Goldman) The Haworth Medical Press, an imprint of The Haworth Press, Inc., 2007, pp. 67-75. Single or multiple copies of this article are available for a fee from The Haworth Document Delivery Service [1-800-HAWORTH, 9:00 a.m. - 5:00 p.m. (EST). E-mail address: docdelivery@haworthpress.com].

Available online at http://ja.haworthpress.com
© 2007 by The Haworth Press, Inc. All rights reserved.
doi:10.1300/J096v12n01_07

terns in the U.S. are significant on a global basis, because the U.S. has such a large share of pesticide use globally–24% of agricultural and household pesticides–in relationship to population.[1] Also, the U.S. is a major exporter of pesticides to the rest of the world.

The management of pesticide risks is a large and complex enterprise. It includes a myriad of activities and approaches, including: right-to-know laws, product labeling, measures to mitigate and monitor pesticides in groundwater, pesticide applicator requirements and training, worker safety regulations (including reentry intervals), pesticide spray restrictions, education and training provided by state extension services, safety engineering and design, watershed and wildlife protection, and farm and migrant farm worker education. All of these are important but this paper will not provide a broad overview of all these areas. Rather, this paper will focus on management of risks to agricultural pesticides that potentially have chronic health risks, such as those discussed in the rest of this monograph. Specifically it is focused on

two tools that can impact on chronic health risks of pesticides. The first of these is the decision to register a pesticide active ingredient and/or specific product formulations. The second is the pesticide tolerance (called a maximum residue limit in other countries) which controls the allowable level of a pesticide on an individual agricultural commodity. Both of these decisions involve a complex risk analysis and here the specific focus is on the assessment of exposures and chronic health risks to humans.

From 1964, the first year for which EPA has data, agricultural and household use of pesticides increased from 600 to an all-time high of 1.1 billion pounds in 1979 (Figure 1). The annual rate of increase during this time was an average of 4.2% a year. By 2001 use had decreased to about 890 million pounds. While usage rates have fluctuated somewhat since 1979, the overall rate of decrease was an average of 1% per year until 2001, the most recent year with data.[1] Much of this decrease is attributable to decreased use for nonagricultural (home and garden) uses. Changes in agricul-

FIGURE 1. Conventional pesticide usage trends, US, millions of pounds active ingredients

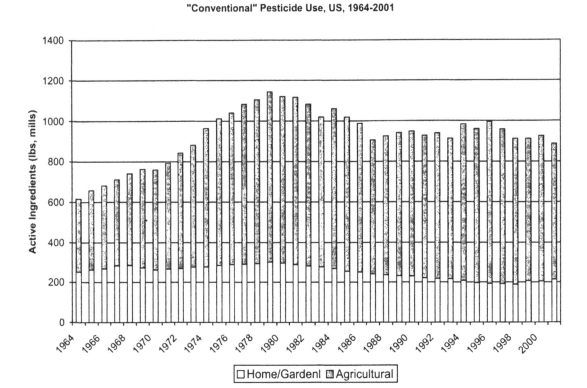

tural use could be attributable to a number of factors including reduction in unnecessary uses, shifts to more potent pesticides in some cases and in other cases shifts away from conventional pesticides to other technologies such as "plant incorporated protectants" (plants genetically modified to have new pest resistance or pesticidal traits), biological controls and so forth.

CHRONOLOGY OF U.S. PESTICIDE LEGISLATION

Congress enacted the first pesticide legislation 1910; this was a consumer protection statute that "aimed to reduce economic exploitation of farmers by manufacturers and distributors of adulterated or ineffective pesticides."[2] Initially, the U.S. Department of Agriculture (USDA) was responsible for administering the pesticide statutes; responsibility shifted to the EPA when it was created in 1970. Pesticides also had been regulated in food under the Federal Food, Drug and Cosmetics Act (FFDCA) for residues in food. In 1947, Congress enacted the Federal Insecticide, Fungicide and Rodenticide Act (FIFRA), which for the first time addressed the potential risks to human health posed by pesticide products.[3]

In 1962, Rachel Carson's book *Silent Spring* brought new public attention to the debate about pesticide risks producing a major shift in emphasis from pesticide regulation as a tool for consumer protection for farmers to pesticide regulation as a tool for protection of the environment.[4] In 1972, Congress had become concerned about pesticide impacts to human health and the environment, because of Carson's book and other signs of environmental and human health impacts from use of pesticides. Congress completely revised FIFRA to strengthen the EPA's authority in a number of regards, including giving EPA the ability to restrict the use of a pesticide.[5] The 1972 law is the basis of current federal policy. FIFRA has a number of provisions that give the EPA the authority to assess and manage the risks of pesticides. These include provisions related to registration of new pesticides, reregistration of older pesticides, data collection, and the ability to restrict pesticide import and export to a certain extent. Since 1972, EPA has been required to reregister older pesticides, and during the 1970s-80s EPA imposed bans or severe restrictions on DDT, aldrin, endrin, chlordane, heptachlor, toxaphene, dibromochloropropane (DBCP), ethylene dibromide (EDB) and other highly chlorinated and brominated pesticides, as well as a number of pesticides that contained toxic metals such as lead, arsenic and thallium. At the same time, several less persistent but highly toxic pesticides, like organophosphates and carbamates, increased in usage during that time.

During the 1980s, in the U.S. concerns continued to mount about pesticide risks. This is a time when chronic health hazards associated with certain pesticides began to be more widely appreciated. Dioxins as a contaminant of the herbicide 2,4,5-T (Agent Orange) famously provoked concern in the context of the Vietnam war, where the herbicide was used as a defoliant by the U.S. military. In the 1970s, DBCP was discovered as a cause of infertility among exposed male pesticide-manufacturing workers.[6] In the 1980s, the pesticide chlordimeform was linked to bladder cancers in manufacturing workers.[7] These episodes, while unusual, helped policy makers understand the connection between theoretical risks of pesticide as determined from animal testing to disease burdens in populations. From the 1960s onward episodes of severe pesticide-related poisonings due to organophosphate pesticides were reported worldwide (including in the U.S.);[8-10] it is less certain what the chronic health risks have been from such exposures, the majority of which have been occupational but also due to accidental ingestions by children or suicide attempts.

Concerns about such chronic health risks, along with the very slow progress of reassessment of older pesticides under the 1972 law, led to new legislative efforts. In 1988 Congress enacted milestone legislation, amending FIFRA to accelerate the process of review and update of older pesticides.[11] The "older" pesticides were those that had come onto the market in the U.S. prior to imposition of new, more stringent, registration standards in 1985. Because they were registered prior to the development of these safety standards, they posed the greatest risks to health and the environment. In 1988, there were 612 such active ingredients on the market. Congress set an aggressive schedule for reregistration that EPA was not able to follow.

At the time Congress required pesticide reregistration, there was particular focus on 10 carcinogenic pesticides: linuron, zineb, captafol, captan, maneb, permethrin, mancozeb, folpet, chlordimeform, and chlorothalonil. The National Research Council in 1985 had reported that these 10 pesticides accounted for 80 to 90 percent of the total estimated dietary cancer-causing risk from the 28 carcinogenic pesticides that were then commonly found on food.[12] By 1995, the GAO found that the EPA had made significant progress toward regulating these pesticides,[13] however, it took a long time for EPA to get the job done. Reassessments for mancozeb and maneb were completed in 2005, 20 years after the NRC report; the reassessment for permethrin was completed a year later.

In the 1990s concerns shifted from an emphasis on carcinogenicity to a focus on risks to children. Many of the basic concepts for protection of children's health in pesticide regulation were first put forward in the 1993 National Research Council (NRC) report, "Pesticides in the Diets of Infants and Children." The NRC concluded that the intrinsic hazards and exposures levels of pesticides are frequently different for children and adults, and that the EPA did not adequately address those differences.[15] The committee advised the EPA to incorporate information about dietary exposures to children in risk assessments and augment pesticide testing with new or improved guidelines for neurotoxicity, developmental toxicity, endocrine effects, immunotoxicity, and developmental neurotoxicity. It recommended that EPA include cumulative risks from pesticides that act via a common mechanism of action and aggregate risks from non-food exposures when developing a tolerance for a pesticide.[15]

FOOD QUALITY PROTECTION ACT OF 1996

Shortly after publication of the 1993 report, the Clinton Administration announced an initiative to address the NRC recommendations, including asking Congress for new legislative authorities. The 1996 Food Quality Protection Act (FQPA) gave the agency one uniform standard to use in setting tolerances, which are the limits of allowable pesticide residues on a food, a "reasonable certainty of no harm."[16]

In establishing tolerances the FQPA requires that the agency consider information on the aggregate of all non-occupational exposures, including drinking water and exposures from lawn and household uses. Although the implementation of this provision may be imperfect, previously, the law allowed EPA only to account for pesticides in food when setting tolerances. The law also requires the EPA to consider available information on cumulative effects of pesticide residues and other substances that have a common mechanism of toxicity. Also, Congress required EPA to use an additional 10-fold (10X) uncertainty factor in risk assessment to account for pre- and post-natal toxicity.[16] Such a third 10X factor was recommended by the National Research Council.[15] According to the law, the agency can eliminate or reduce this additional 10X FQPA factor for children only if it makes a finding that reliable and complete data indicate a different factor will be safe for infants and children. Specifically, the FQPA instructed EPA:

> In the case of threshold effects, . . . an additional tenfold margin of safety for the pesticide chemical residue and other sources of exposure shall be applied for infants and children to take into account potential pre- and post-natal toxicity and completeness of the data with respect to exposure and toxicity to infants and children. Notwithstanding such requirement for an additional margin of safety, the Administrator may use a different margin of safety for the pesticide chemical residue only if, on the basis of reliable data, such a margin will be safe for infants and children.[16]

Since the enactment of the FQPA, there has been much debate about the implementation of these new provisions. To date, the most successful has been the aggregate risk policy.[17] A number of household pesticide registrations were revoked when it was discovered by the EPA that they led to an excessive level of exposure in children. These include household uses of chlorpyrifos, malathion, and diazinon.

The 10X FQPA factor has been controversial; both industry and environmental groups have argued that EPA has applied it inappropriately. The herbicide bromoxynil was a case that raised concerns for industry. A major assessment of bromoxynil was occurring at the same time that FQPA was enacted and this created a complex situation for EPA decision makers. The most sensitive endpoint identified for bromoxynil was developmental toxicity, namely supernumerary ribs in test animals. As stated in the 1998 registration eligibility document: "Upon review of the extensive developmental toxicological database for this chemical, a concern for in utero developmental effects was noted. In order to provide a sufficient margin of safety for the developing fetus, the 10-fold safety factor for enhanced sensitivity to infants and children was retained for females 13 + thus requiring a 1000-fold uncertainty factor for this population subgroup." [United States Environmental Protection Agency. Office of Pesticide Programs, 1998 #5080]. What the document did not state is that prior to the enactment and implementation of FQPA, the agency had been on a course to assess the bromoxynil hazards with a 100-fold factor, in other words, not to include an additional factor to protect children. In my role at the agency at the time, I was faced with a split between the scientists who reviewed the data. Some of EPA's experts pointed to the steep dose response curve for developmental effects of bromoxynil (which progressed from skeletal abnormalities to fetal death within a fairly narrow dose range). Others felt that since the hazards were well characterized, no additional uncertainty factor would be required under the FQPA. It was my decision to make and I opted for the more protective 1000-fold factor. As a result, in January 1998, EPA denied a petition for a new use of bromoxynil on cotton genetically modified to be tolerant to this herbicide. At that time the cotton industry was prepared to plant millions of acres of bromoxynil tolerant cotton in the next growing season. The denial of the bromoxynil tolerance on cotton generated a political firestorm by the pesticide industry and some farm groups against the implementation of FQPA.[18]

EPA developed an "FQPA factor" policy, which provides guidance for how EPA will determine that a different factor is appropriate. Over the years, EPA has struggled with how to best incorporate this FQPA factor. At the time that FQPA was enacted, I was Assistant Administrator for EPA's Office of Prevention, Pesticides and Toxic Substances and was asked to assess whether the new law would really change how EPA assesses hazards to children. At the time, EPA's science staff advised that it would not, because EPA applies a "data base uncertainty factor" of 10 for noncancer health hazard assessment when it lacks data relevant to the health of children. However, once FQPA was enacted, there was a major change. With a legal requirement to make a finding that children's hazards and exposures were addressed directly or via the incorporation of an additional uncertainty factor, the FQPA tolerance reassessment resulted in the elimination of numerous uses of pesticides, especially organophosphate pesticides, despite the fact that many scientists at EPA felt that they already were assessing risks to children.

Environmental groups have argued that EPA applies the FQPA 10X factor too infrequently.[19] Table 1 shows how the FQPA uncertainty factor was applied from the time of enactment of the law through 1999, for pesticides in general and for organophosphates in particular.[17] Of the first 105 pesticides reviewed, EPA decided not to retain any portion of the additional 10X factor for 56, to retain a full 10X factor for 21, and to retain a partial 3X factor for 23. In five cases EPA determined that the FQPA 10X uncertainty factor should be greater than 10. However, for the 39 organophosphates, a 10X (or more) factor was retained for 12 pesticides and partly retained (3X) for another 12 pesticides.[17]

COMPLETING THE REASSESSMENT OF OLDER PESTICIDES

Pesticide Reregistration: By the end of 2005, EPA reported that reregistration had been completed for 271 of these pesticides and that 231 had been canceled; EPA projects that the remaining 110 pesticides scheduled for reregistration would be completed by October 2008.[20,21] Although this well behind the sched-

TABLE 1. EPA application of additional FQPA factors (to assure safety to children): 1996-2000.[17]

Type of pesticide reviewed	Number reviewed	No additional factor	3-fold factor	10-fold Factor	>10 fold factor
Organophosphates	39	15	12	10	2
All Others	66	41	11	11	3
Total	105	56	23	21	5
Percentage		53%	22%	20%	5%

ule envisioned by Congress, the process has been relatively successful and, it is certainly the case that pesticide reregistration is a complex process that requires much effort by EPA's science staff (as well as by the pesticide industry). For example, EPA received 27,260 scientific studies to support reregistration of pesticides in this program.[14]

Tolerance Reassessment: The FQPA directed EPA to reassess all 9,721 existing tolerances over 10 years; the deadline for completion of this task was August 2006. The Act required that EPA establish a schedule to assess the pesticides with greatest potential risk first. EPA found that a large proportion, 57% of tolerances, fell into the highest risk category. These include probable/possible carcinogens (20.7%) and organophosphate pesticides (17.4%). As of August 2006 EPA reported that there were 94 tolerances remaining to be reassessed; 3,200 tolerances were revoked and 1,200 were modified.[20-22] To conduct this reassessment required consideration of "cumulative risk" to pesticides that act via a common mode of action. EPA determined that there were four groups of pesticide active ingredients that should be assessed in that manner: organophosphate insecticides, carbamate insecticides, triazine herbicides and chloracetanilide herbicides. All of these are of concern for chronic health endpoints: The organophosphate and carbamate insecticides particularly have been associated with chronic as well as acute neurotoxic effects. The triazines–atrazine, simazine and propazine, as well as three of their degradants–may disrupt the hypothalamo-pituitary axis and thus may have endocrine effects.[23] EPA also concluded that three of the chloroacetanilides, acetochlor, alachlor, and butachlor–should be assessed cumulatively not only on the basis that all three are carcinogens but also because they cause nasal turbinate tumors in which are animal carcinogens that cause similar tumors in lab animals.[24] At this time, the EPA has concluded that the cumulative risk from allowable uses of these four groups of pesticides are acceptable.[22]

Outlook for the Future

In consequence of these efforts, are pesticide exposures–and risks–decreasing? For one group of pesticides that have largely been assessed, the organophosphates, there is little evidence for change in usage patterns, based on the sales data that are compiled by the EPA (which only are through 2001). As is shown in Figure 2, there is little change in either total insecticide use or use of organophosphate pesticides through 2001.[1] However, other more recent data support a general decline in exposure to these pesticides. The CDC Third National Report on Human Exposure to Environmental Chemicals reported that there were significant declines in average and "high end" (90th and 95th percentile) levels of three of six organophosphate metabolites in urine that were monitored both in 1999-2000 and 2001-2002. Two metabolites generally were below the level of detection in both periods and one, diethylthiophosphate, showed a small increase in the median level and no change in the "high end" levels. 3,5,6-Trichloro-2-pyridinol (TCPy), a specific metabolite of chlorpyrifos and chlorpyrifos methyl, showed a similar trend.[25] Investigators in New York City reported decreases in household air and maternal plasma samples for several pesticides between 1998-2001, including chlorpyrifos, diazinon, propoxur, and bendiocarb.[26]

Other trends that are favorable that are reported by the EPA are the banning of most

FIGURE 2. Trends in insecticide use: total insecticides and organophosphates, U.S., millions of pounds

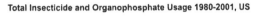

Total Insecticide and Organophosphate Usage 1980-2001, US

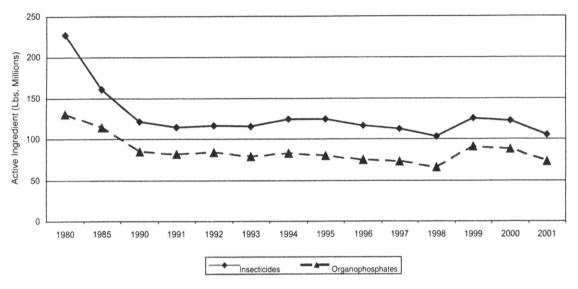

household uses of organophosphate, carbamate and other acutely toxic pesticides, sharp increases over the last 20 years in registrations of new pesticides that are biologicals or "reduced risk" and a decline in new registrations for traditional, more highly toxic, types of pesticides.[21] I would be remiss if I did not at least mention the remarkable growth in the U.S. of organic farming, which does not use chemical pesticides; growth was estimated to be around 20% annually over 15 years, reaching an estimated 1.8% of the U.S. food market in 2003. Although these trends are modest, they are encouraging.[27,28]

NEW PESTICIDE REGISTRATIONS

Registration of new pesticides plays an important role in assuring that pesticide users have safer alternatives to older, more hazardous pesticides. In addition, it prevents the introduction on the market of new pesticides that would cause damage in the future. Two things had changed by 1994. In 1993, the EPA established a "reduced risk" pesticide registration process for chemical pesticides.[29] If a pesticide manufacturer could show that a new pesticide attains a "safer" threshold, based on a number of health

and environmental considerations, EPA accelerates the review, thus providing an incentive to manufacturers. In 1994, the EPA established a policy whereby new biological pesticides would be a priority for registration and established a new pesticide review division specifically to accelerate their review. Later, the FQPA included incentives for safer pesticides as well and these were reflected in 1997 policy guidance issued by EPA.[30] Over time, the results of these new policies were evident. Of 125 new pesticides that were approved between 1984-93, 75% were "conventional" pesticides; for the 300 approved between 1994-2004, only 41% were "conventional" pesticides and in the last year 2004, this was down to 21%. Twenty-one percent were new "reduced risk" chemical pesticides and 58% were biological pesticides.[20]

It would appear that incentives for industry to bring forward new, safer chemical and biological pesticides are beginning to achieve success. What this demonstrates is a willingness of industry to develop new and safer products, if given appropriate market incentives in the regulation of pesticides. What is unknown at this time is the extent to which these newer safer pesticides are being widely adopted by pesticide users, and reducing human exposures to more hazardous pesticides.

VOLUNTARY EFFORTS

Numerous initiatives have been undertaken by the EPA in the last several years to reduce risks of existing pesticides, including specific efforts to protect farm workers[21,31] and ground water, and to provide the public with clearer information on pesticide product labels.[32] Many of these have been voluntary efforts directed to pesticide users. In the area of integrated pest management (IPM), the EPA developed an educational program on how to adopt IPM in schools. More broadly, the EPA developed the "Pesticide Environmental Stewardship Program" or PESP, which enlisted dozens of pesticide users (ranging from the U.S. military to various grower groups) who work with the EPA to identify strategies to reduce the risk and use of pesticides in their operations. The USDA in 1993 adopted a goal of 75% of U.S. agriculture using IPM by the year 2000. Whereas the USDA in 2001 reported that this goal nearly was reached (with 70% of agriculture in IPM), according to the GAO this was achieved by the USDA's inclusion of a number of farm management practices (cleaning equipment and monitoring for pests) that did not transform the use of pest control agents in agriculture.[33] For example, they note for one crop "biologically-based IPM practices were implemented on no more than 18 percent of corn acreage (p. 2)." The GAO concluded that the USDA IPM initiative lacks leadership, coordination and management. It recommended strengthening the management and coordination of USDA's IPM effort, as well as more clearly articulating and measuring progress toward achieving the goals (including the goal of pesticide use reduction). An additional problem is that there is no detailed information on pesticide usage on a national basis. Several states (California, New York, Washington, Arizona, and Texas) do collect such data (mostly at the point of sale) and could serve as models for development of a national reporting system.

CONCLUSIONS

The pesticide law in the U.S. has for years given the EPA much authority to gather information about pesticides and to manage their risks. Today the most hazardous pesticides in commerce in the U.S. are organophosphates, the remaining organochlorines, and a number of pesticides that have been classified as carcinogens. The U.S. shows some signs of beginning a downturn in overall use of pesticides, as well as use of some of the most hazardous pesticides. It remains to be seen whether the strong precautionary and child protective provisions of the pesticide legislation (cumulative and aggregate risk assessment and increased safety for children) will be effective in actually reducing the exposures (and risks) of pesticides to workers and the general public.

REFERENCES

1. Kiely T, David D, Grube AH. Pesticide Industry Sales and Usage: 2000 and 2001 Market Estimates. Report No. 733-R-04-001. Available from: *http://www.epa.gov/oppbead1/pestsales/01pestsales/market_estimates2001.pdf* [cited 2007 Feb. 7].

2. U.S. Code. Insecticide Act of 1910. Ch 191, 36 Stat. 331 (7 U.S.C. 121 et seq.).

3. U.S. Code. Federal Insecticide, Fungicide, and Rodenticide Act Amendments, 1947. Ch.125, 86 Stat. 973 (Pub. L. 92-516).

4. Carson R. Silent spring. Boston, MA: Houghton Mifflin; 1962.

5. U.S. Code. Federal Environmental Pesticide Control Act of 1972. 86 Stat. 973 (Pub. L. 92-516).

6. Whorton MD, Wong O, Morgan RW, Gordon N. An epidemiologic investigation of birth outcomes in relation to dibromochloropropane contamination in drinking water in Fresno County, California, USA. Int Arch Occup Environ Health. 1989;61(6):403-7.

7. Popp W, Schmieding W, Speck M, Vahrenholz C, Norpoth K. Incidence of bladder cancer in a cohort of workers exposed to 4-chloro-o-toluidine while synthesising chlordimeform. Br J Ind Med. 1992;49(8):529-31.

8. Kahn E. Epidemiology of field re-entry poisoning. J Environ Pathol Toxicol. 1980;4(5-6):323-30.

9. Jeyaratnam J. Acute pesticide poisoning: a major global health problem. World Health Stat Q. 1990;43(3): 139-44.

10. Blondell J. Epidemiology of pesticide poisonings in the United States, with special reference to occupational cases. Occup Med. 1997;12(2):209-20.

11. U.S. Code. Federal Insecticide, Fungicide, and Rodenticide Act Amendments, 1988. 102 Stat. 2654 (Pub. L. 100-532).

12. National Research Council. Regulating Pesticides in Food: the Delaney Paradox. Washington, DC: National Academies Press; 1987. Available from: *http://*

books.nap.edu/books/0309037468/html [cited 2007 Feb 7].

13. General Accounting Office. Reregistration Status of 10 Highest-risk Dietary Cancer-Causing Pesticides. Washington, DC: GAO, 1995. Report No. B-265923. Available from: *http://archive.gao.gov/paprpdf1/155332.pdf* [cited 2007 Feb 7].

14. Environmental Protection Agency. Pesticide Reregistration Performance Measures and Goals. Federal Register 2004;69(87):25082-25092. Available from: *http://www.epa.gov/EPA-PEST/2004/May/Day-05/p10213. htm* [cited 2007 Feb 5].

15. National Research Council. Pesticides in the Diets of Infants and Children. Washington, DC: National Academies Press; 1993. Available from: *http://newton. nap.edu/books/0309048753/html/R1.html* [cited 2007 Feb 6].

16. U.S. Code. Food Quality Protection Act, 1996. 110 Stat. 1489 (P.L. 104-170).

17. General Accounting Office. Children and Pesticides: New Approach to Considering Risk Is Partly in Place. Washington, DC: GAP; 2000 September 2000. Report No.: GAO/HEHS-00-175. Available from: *http:// www.gao.gov/new.items/he00175.pdf* [cited 2007 Feb 7].

18. EPA Denies Bromoxynil Tolerances On Cotton; Company to Challenge Decision. Daily Environment Report 1998 January 6, 1998:A-1-2.

19. McGarity TO. Politics by other means: law, science, and policy in EPA's implementation of the food quality protection act. Administrative Law Review 2001;53(103).

20. Environmental Protection Agency. Taking Care of Business: Protecting Public Health and the Environment. EPA's Pesticide Program FY 2004 Annual Report. Washington, DC: EPA; 2005. Report No.: EPA 735-R-05-001. Available from: *http://www.epa.gov/ oppfead1/annual/2004/04annualrpt.pdf.* [cited 2007 Feb. 6].

21. Environmental Protection Agency. EPA's Pesticide Program. FY 2005 Annual Report. Washington, DC: EPA; 2006. Available from: *http://www.epa.gov/ oppfead1/annual/2005/05annualrpt.pdf* [cited 2007 Feb. 5].

22. Environmental Protection Agency. Accomplishments under the Food Quality Protection Act (FQPA) August 3, 2006 – 10th Anniversary of the Food Quality Protection Act. Washington, DC: EPA Office of Pesticide Programs; 2006 August 3, 2006. Available from: *http://www.epa.gov/pesticides/regulating/laws/fqpa/fqpa_ accomplishments.htm* [cited 2007 Feb. 7].

23. Environmental Protection Agency. The grouping of a series of triazine pesticides based on a common mechanism of toxicity. Washington, DC: U.S. Environmental Protection Agency; 2002.

24. Mulkey ME. A Common Mechanism of Toxicity Determination for Chloroacetanilide Pesticides (Memorandum to Lois Rossi, Director, Special Review and Reregistration Division and to Jim Jones, Director, Registration Division). Memorandum. Washington, DC: US EPA Office of Pesticide Programs; 2001 July 10, 2001.

25. Centers for Disease Control and Prevention. Third National Report on Human Exposure to Environmental Chemicals. Atlanta: CDC; 2005 2005. Available from: *http://www.cdc.gov/exposurereport/3rd/pdf/ thirdreport.pdf* [cited 2007 Feb. 5].

26. Whyatt RM, Barr DB, Camann DE, Kinney PL, Barr JR, Andrews HF, Hoepner LA, Garfinkel R, Hazi Y, Reyes A, Ramirez J, Cosme Y, Perera FP. Contemporary-use pesticides in personal air samples during pregnancy and blood samples at delivery among urban minority mothers and newborns. Environ Health Perspect. 2003;111(5):749-56.

27. Thilmany D. The US organic industry: important trends and emerging issues for the USDA. Agribusiness Marketing Report 06-01, Colorado State University. Available from: *http://asap.sustainability.uiuc.edu/ members/dananderson/wortfolder/abmr06-01.pdf/view* [cited 2007 Feb. 6].

28. Oberholtzer L, Dimitri C, Greene C. Price premiums hold on as organic produce market expands. Washington, DC: USDA Economic Research Service; 2005. Report No.: Outlook Report VGS-308-01. Available at: *http://www.ers.usda.gov/publications/vgs/may05/ VGS30801/VGS30801.pdf* [cited 2007 Feb. 5].

29. Environmental Protection Agency. Voluntary Reduced-Risk Pesticides Initiative. Washington, DC: EPA; 1993 July 21, 1993. Report No.: PR Notice 93-9.

30. Environmental Protection Agency. Guidelines for Expedited Review of Conventional Pesticides under the Reduced-Risk Initiative and for Biological Pesticides. Washington, DC: EPA; 1997 September 4, 1997. Report No.: PR 97-3. Available at: *http://www.epa.gov/ opppmsd1/PR_Notices/pr97-3.html* [cited 2007 Feb. 7].

31. Environmental Protection Agency. Worker Protection Standard. Federal Register 1992;57(Aug. 21, 1992):38151. Available at: *http://www.epa.gov/pesticides/ safety/workers/PART170.htm* [cited 2007 Feb. 7].

32. Abt Associates Inc. Consumer Labeling Initiative: Phase II Report. Cambridge, MA: Abt Associates Inc.; 1999 October, 1999. Report No.: EPA Contract Number: 68-W6-0021. Available from: *http://www. p2pays.org/ref/17/16262.pdf* [cited 2007 Feb. 5].

33. General Accounting Office. Agricultural Pesticides: Management Improvements Needed to Further Promote Integrated Pest Management. Washington, DC: GAO; 2001 August 2001. Report No.: GAO-01-815. Available from: *http://www.gao.gov/new.items/d01815. pdf* [cited 2007 Feb. 7].

doi:10.1300/J096v12n01_07

Appendix A:
Informational Sources on Pesticides and Health

Daniel L. Sudakin, MD, MPH, FACMT, FACOEM

Source	Web Address (URL)	Type of Information
United States Environmental Protection Agency (US EPA) Type of resource: P	http://www.epa.gov/oppfead1/labeling/lrm/ individual chapters of the Pesticide Label Review Manual, 3rd Edition	Pesticide Label Review Manual, 3rd Edition. Utilized by the EPA Office of Pesticide Programs (OPP) to improve the quality and consistency of pesticide labels. This document provides guidance on how the US EPA interprets the Federal Insecticide, Fungicide, and Rodenticide Act (FIFRA) and its specific requirements for label language and format. Chapters include: • Precautionary labeling (including Signal Word, Personal Protective Equipment, and First Aid Statements) • Physical or Chemical Hazards • Worker Protection Labeling • Directions for Use • Labeling Claims • Storage and Disposal • Identification numbers (including EPA registration number)
US EPA Type of resource: P, E	http://www.epa.gov/opptintr/labeling/ http://www.epa.gov/opptintr/labeling/pubs/ v1q39loi.pdf Frequently Asked Questions http://www.epa.gov/oppfead1/pmreg/ Pesticide Management Resource Guide (PMReG)	Consumer Labeling Initiative: A voluntary effort of EPA, industry, other federal and state agencies, and private groups, aiming to make health, safety, and environmental information on household pesticide products easier to find, understand, and use. The Frequently Asked Questions section describes interim label changes that are being implemented including: • Telephone numbers on product labels • Common names, instead of formal chemical names • Use of term "other ingredients" instead of "inert ingredients" • Use of term "first aid" instead of "statement of practical treatment." The Pesticide Management Resource Guide provides information resources at EPA and elsewhere. The information is intended for use in pesticide management decision-making.

Daniel L. Sudakin is Associate Professor, Department of Environmental and Molecular Toxicology, Oregon State University, Corvallis, OR.

Address correspondence to: Daniel L. Sudakin, MD, MPH, Department of Environmental and Molecular Toxicology, Oregon State University, 333 Weniger, Corvallis, OR 97331 (E-mail: sudakind@ace.orst.edu).

[Haworth co-indexing entry note]: "Appendix A: Informational Sources on Pesticides and Health." Sudakin, Daniel L. Co-published simultaneously in *Journal of Agromedicine* (The Haworth Medical Press, an imprint of The Haworth Press, Inc.) Vol. 12, No. 1, 2007, pp. 77-82; and: *Proceedings from the Medical Workshop on Pesticide-Related Illnesses from the International Conference on Pesticide Exposure and Health* (ed: Ana Maria Osorio, and Lynn R. Goldman) The Haworth Medical Press, an imprint of The Haworth Press, Inc., 2007, pp. 77-82. Single or multiple copies of this article are available for a fee from The Haworth Document Delivery Service [1-800-HAWORTH, 9:00 a.m. - 5:00 p.m. (EST). E-mail address: docdelivery@haworthpress.com].

Available online at http://ja.haworthpress.com
© 2007 by The Haworth Press, Inc. All rights reserved.

doi:10.1300/J096v12n01_08

Source	Web Address (URL)	Type of Information
California Department of Pesticide Regulation (DPR), and US EPA Type of resource: HA	*http://oaspub.epa.gov/pestlabl/ppls.home* US EPA Pesticide Product Label System (PPLS) *http://www.epa.gov/pesticides/pestlabels/disclaimer.htm* Limitations of the EPA PPLS	Searchable database of registered and cancelled pesticide products in the United States. The database includes: • Product names and EPA registration numbers • Company names and telephone numbers • Registration and/or cancellation dates • EPA product manager name and phone number The databases can be searched by PC codes, product names, firm number or company name, as well as EPA registration number. The search results are linked to the EPA Pesticide Product Label System (PPLS), which includes TIFF images of product labels.
Crop Data Management Systems, Inc. Type of resource: HA	*http://www.cdms.net/ACM.aspx* Pesticide manufacturer list *http://www.cdms.net/LabelsMsds/LMDefault.aspx* Search engine for pesticide labels and Material Safety Data Sheets	Website contains a list of pesticide manufacturers, with linkages to websites where additional information about their products can be accessed (including labels and Material Safety Data Sheets). Searches can be performed by pesticide product name, product type, crop, and pest.
Center for Environmental and Regulatory Information Systems (CERIS), Purdue University Type of resource: HA, RA, P	*http://ppis.ceris.purdue.edu/index.html*	The National Pesticide Information Retrieval System (NPIRS) is a collection of pesticide-related databases that is accessible by paid subscription. The database includes: • Pesticide product label information (federal and state) • Documentation in support of pesticide product information submitted to US EPA • Pesticide tolerances • Federal Register articles from Environmental and Agriculture groups from 1993-present
EXTOXNET Type of resource: HA, EA, DR, RA	*http://extoxnet.orst.edu/ghindex.html*	A cooperative effort of the University of California-Davis, Michigan State University, Cornell University, and University of Idaho. Contains documents providing pesticide information relating to health and environmental effects of specific active ingredients. Information includes: • Trade names • Regulatory status • Chemical class • Formulations • Toxicological effects • Ecological effects • Environmental fate • Physical properties • Exposure guidelines • Scientific references

Source	Web Address (URL)	Type of Information
National Pesticide Information Center (NPIC) Type of resource: HA, EA, DR, RA, P, E	*http://www.npic.orst.edu/npicfact.htm* Fact sheets on pesticide topics and active ingredients *http://www.npic.orst.edu/state1.htm* Links to state pesticide regulatory agencies *http://www.npic.orst.edu/rmpp.htm* Electronic version of "Recognition and Management of Pesticide Poisonings, 5th Ed." *http://www.npic.orst.edu/mcapro/index.html* NPIC medical case profiles for health care providers	A cooperative effort of Oregon State University and US EPA. Provides general and technical fact sheets on pesticides and active ingredients. Fact sheets include: • Pesticide classification • Uses • Mechanism of action • Acute and chronic toxicity • Metabolism • Environmental fate • Effects on wildlife Pesticide topic fact sheets include: • Inert ingredients • Pesticide formulations • Pesticides in indoor air • Pesticides in drinking water • Pets and pesticide use • Signal words • Wildlife and pesticides Additional resources include: • Links and contact information to State Lead Agencies for Pesticide Regulations • Electronic version of *Recognition and Management of Pesticide Poisoning, 5th Ed.* • Medical case profiles (for health care providers)
California DPR Type of resource: DR	*http://www.cdpr.ca.gov/docs/toxsums/toxsumlist.htm*	Pesticide Registration Branch Toxicology Data Review Summaries. Includes technical summaries on pesticidal active ingredients.
California Environmental Protection Agency Type of resource: DR, P	*http://www.oehha.ca.gov/pesticides/pdf/docguide2002.pdf*	Guidelines for physicians who supervise workers exposed to cholinesterase-inhibiting pesticides (Fourth Edition, 2002). Includes chapters on: • Occupational health services • Pre-exposure examinations • Cholinesterase monitoring • Cholinesterase testing • Permissible levels of cholinesterase depression • Removal from exposure and return to work • Frequency of periodic follow-up testing • Prophylaxis, medical treatment, and first aid • Physician's responsibility to report pesticide-related illness and injury
Washington State Department of Labor and Industries Type of resource: DR, P, E	*http://www.lni.wa.gov/Safety/Topics/AtoZ/Cholinesterase/Providers.asp*	Cholinesterase monitoring information for medical providers. Website includes description of medical monitoring and laboratory testing for cholinesterase activity. Includes links to: • Cholinesterase monitoring fact sheets for employees (English and Spanish) • Guidelines for health care providers • Sample cholinesterase medical history questionnaires

Source	Web Address (URL)	Type of Information
Agency for Toxic Substances and Disease Registry (ATSDR) Type of resource: HA, EA, DR, RA, P, E	*http://www.atsdr.cdc.gov/toxpro2.html*	ATSDR Toxicological Profiles. Includes profiles for numerous pesticidal active ingredients, including pesticides that are no longer registered for use in the United States. Profiles include: • Public health statement • Health effects data • Chemical and physical information • Production, import, use, and disposal data • Human exposure assessment • Analytical methods • Regulations and advisories • Scientific references
Centers for Disease Control and Prevention (CDC) Type of resource: HA, EA, E	*http://www.cdc.gov/exposurereport/*	National Report on Human Exposure to Environmental Chemicals. A longitudinal assessment of the U.S. population exposure to environmental chemicals. The Third Report includes biomonitoring data on human exposure to: • Organochlorine insecticides • Pyrethroid insecticides • Organophosphate insecticides • Herbicides • Insect repellents (DEET)
National Environmental Education and Training Foundation (NEETF) Type of resource: P, E	*http://www.neetf.org/health/pesticides/index. htm* *http://www.neetf.org/Health/pestlibrary.htm* NEETF pesticide resource library	NEETF National Strategies for Health Care Providers: Pesticides Initiative. Describes a strategic approach for incorporating environmental health information into the education and practice of health care providers, using pesticides as a model. Includes links to: • Pesticide competency and practice skill guidelines for medical and nursing education and clinical practice • Enviornmental history taking tool • Pesticide resource library
National Library of Medicine (NLM) Type of resource: HA, EA, RA	*http://toxnet.nlm.nih.gov/index.html*	NLM TOXNET database. Searchable databases include: • Hazardous Substances Data Bank (HSDB) • Integrated Risk Information System (IRIS) • International Toxicity Estimates for Risk • Genetic Toxicology (Gene-Tox) • Chemical Carcinogenesis Research Info System (CCRIS) • Toxline • Developmental and Reproductive Toxicity (DART) • Toxics Release Inventory (TRI) • ChemID*plus*
National Institute of Occupational Safety and Health (NIOSH) Type of resource: HA, EA, P, E	*http://www.cdc.gov/niosh/topics/pesticides/*	NIOSH webpage for pesticide illness and injury surveillance. Includes links to: • NIOSHTIC-2 bibliographic database, and description of the Sentinel Event Notification System for Occupational Risks (SENSOR). • articles in peer-reviewed journals and the Morbidity and Mortality Weekly Report (MMWR) on the epidemiology and clinical toxicology of pesticides. • State-based surveillance programs for pesticides

Source	Web Address (URL)	Type of Information
NIOSH Agricultural Centers Type of resource: E	http://www.cdc.gov/niosh/agctrhom.html	A NIOSH effort to protect the health and safety of agricultural workers and their families by conducting research and educational outreach for the prevention of illness and injury. Includes links to NIOSH Agricultural Centers throughout the United States.
Pesticide Action Network North America Type of resource: P, E	http://www.pesticideinfo.org/ Search_Poisoning.jsp#Reporting	A searchable database (by state and county) with information on pesticide exposure incident reporting
National Pesticide Medical Monitoring Program Type of resource: E	http://oregonstate.edu/npmmp/	A cooperative agreement between US EPA and Oregon State University. Provides and collects information on the clinical toxicology of pesticides. Website includes links to pesticide toxicology resources for health care providers.
Association of Occupational and Environmental Clinics Type of resource: HA, EA, RA, E	http://www.aoec.org/	Provides educational resources, epidemiology tools, and a clinic directory for health care professionals specializing in occupational and environmental medicine. Educational resources include powerpoint presentations entitled: • Pesticide Toxicology, Illness Epidemiology, Diagnosis, and Treatment • Fumigants, Fungicides, Herbicides, Disinfectants • Chronic Health Effects, Laws and Regulations
Migrant Clinicians Network Type of resource: E	http://www.migrantclinician.org/	Provides support, technical assistance, and professional development services to clinicians who provide medical care to migrants and other mobile poor populations. Website includes links to resources, including documents in Spanish which focus on prevention of pesticide exposures among migrant workers and their families.
US EPA, Office of Pesticide Programs Type of resource: RA, P, E	http://www.epa.gov/pesticides/health/ worker.htm	US EPA Web page on worker safety and training. Includes links to: • Worker Protection Standard (WPS) • Certification and training regulations • Laboratories approved for cholinesterase testing (occupational surveillance) • Restricted use products report • Pesticide safety programs • National dialogue on the WPS • Resources on worker protection issues
Agricultural Health Study (National Institutes of Health, National Institute of Environmental Health Sciences, National Cancer Institute, US EPA) Type of resource: EA, E	http://www.aghealth.org/	An epidemiological study of pesticides and human health, including private and commercial pesticide applicators. Includes link to: • Study background • Frequently asked questions • Important findings from the study • Publications • Other resources for agricultural health information • Information for scientific collaborators

Source	Web Address (URL)	Type of Information
Virginia Polytechnic Institute and State University Type of resource: E	*http://scholar.lib.vt.edu/ejournals/JPSE/*	Journal of Pesticide Safety Education, the official journal of the American Association of Pesticide Safety Educators. Publishes articles on instructional methods, training devices, research findings, and publication reviews.
Purdue University Extension Service Type of resource: E	*http://www.btny.purdue.edu/ppp/PPP_pubs. html*	Purdue pesticide programs fact sheets. Includes links to fact sheets oriented towards consumers on topics including: • Pesticides and the label • Pesticides and their proper storage • Pesticides and spill management • Pesticides and the community right-to-know • Pesticides and material safety data sheets • Pesticides and personal protective equipment

Appendix B:
Pesticide Intoxication Reporting Forms

Nida Besbelli, PhD

INTRODUCTION

Pesticides are chemicals that require special attention because of their inherent properties that may cause harm to humans and the environment. Although the majority of countries have legislation for the control of pesticides, concerns still exist with respect to their hazards and risks. Pesticide poisoning events are an especially serious health issue in developing countries. Governments, as well as regional and international non-governmental organizations, collect information on acute pesticide intoxications to better understand health problems caused by these compounds, their magnitude and to enable necessary preventive measures.

OBJECTIVES OF COLLECTING INFORMATION ON PESTICIDE INTOXICATIONS

Data recording and reporting of pesticide intoxications are useful to:

- determine the extent of pesticide poisoning incidents and deaths (the size and characteristics of the problem)
- characterize poisoning cases with respect to age, sex, agent, place, circumstances of exposure, estimated risk, and course of treatment
- identify populations at risk
- identify formulations and/or active ingredients causing problems
- find out high risk circumstances associated with workers and pesticide users
- plan and implement prevention and control measures
- provide education and support for physicians and other health care personnel to comply with international conventions, such as the Rotterdam Convention.

The Rotterdam Convention on Prior Informed Consent provides countries with the tools and information needed to identify potentially hazardous chemicals (including pesticides), and to exclude from import those they cannot manage safely. The two key operational elements of the Convention include the prior informed consent procedure, and a process for information exchange among member nations on potentially hazardous chemicals, including some pesticides.[1]

Nida Besbelli is Technical Officer, Chemical Safety, World Health Organization (WHO) European Centre for Environment and Health, Bonn, Germany, WHO Regional Office for Europe.

Address correspondence to: Dr. Nida Besbelli, WHO European Centre for Environment and Health, Bonn, Hermann-Ehlers-Str. 10, D53113 Bonn, Germany.

[Haworth co-indexing entry note]: "Appendix B: Pesticide Intoxication Reporting Forms." Besbelli, Nida. Co-published simultaneously in *Journal of Agromedicine* (The Haworth Medical Press, an imprint of The Haworth Press, Inc.) Vol. 12, No. 1, 2007, pp. 83-90; and: *Proceedings from the Medical Workshop on Pesticide-Related Illnesses from the International Conference on Pesticide Exposure and Health* (ed: Ana Maria Osorio, and Lynn R. Goldman) The Haworth Medical Press, an imprint of The Haworth Press, Inc., 2007, pp. 83-90. Single or multiple copies of this article are available for a fee from The Haworth Document Delivery Service [1-800-HAWORTH, 9:00 a.m. - 5:00 p.m. (EST). E-mail address: docdelivery@haworthpress.com].

Available online at http://ja.haworthpress.com
doi:10.1300/J096v12n01_09

SOURCES OF THE INFORMATION ON PESTICIDE INTOXICATIONS

Data on pesticide intoxications are available from different sources, but the extent of information collected would be different and mostly dependent on the severity and outcome of the intoxication. Death certificates and autopsy/pathology reports can reveal useful information for estimating the incidence of fatal intoxications that would not reach the hospitalized population. Hospital in-patient, emergency room, ambulance and emergency medical technician records as well as poison center records are good sources of information on incidents of severe non-fatal intoxications. A smaller proportion of mild and moderate incidents and exposures would also be collected at poison centers. Although occupational pesticide exposures and intoxications should be recorded at workplaces, most workplaces in developing countries do not maintain records of exposures and intoxications; this is particularly true in the agricultural sector.

Surveillance is the ongoing systematic collection, collation, analysis and interpretation of data; and the dissemination of information to those who need to know in order that action may be taken. Surveillance data can be used to identify pesticide problems, estimate the magnitude of the pesticide poisoning, and evaluate intervention and prevention efforts. Surveillance systems on acute pesticide intoxications would be the best way of assessing the problem of pesticide intoxications, but only a very few countries, mostly in the developed world, have established such systems. In the United States of America, 12 states conduct pesticide illness and injury surveillance. There are some notable examples of pesticide illness surveillance efforts in developing countries as well:

- Through the "Occupational and Environmental Aspects of Exposure to Pesticides in the Central America Isthmus" (PLAGSALUD) Project, surveillance of acute pesticide poisoning was incorporated in the surveillance systems of seven countries (Belize, Costa Rica, El Salvador, Guatemala, Honduras, Nicaragua, Panama).[2]

- Chile established the "Network on Epidemiological Surveillance of Pesticides" (REVEP) in 1999, at the Epidemiology Department of the Public Health Division. With the Decree 88, notification of pesticide poisonings became obligatory in this country in May 2004.[3]
- In Thailand, the Ministry of Public Health conducts nationwide screening of people in the agricultural sector and the disease surveillance system, which includes pesticide poisoning. The results are published in the Annual Epidemiology Surveillance Report.[4]

COMPONENTS OF A PESTICIDE INTOXICATION REPORTING FORM

How much and what type of data are to be collected will depend upon the objectives, priorities, type of organization collecting the data, as well as available resources. However, it is desirable that data collected through the pesticide intoxication reporting forms include the following as a minimum.

1. Communication details: information on the name and job title of the person recording the incident, institution; name and job title of the person requesting and/or treating the patient, institution, contact details, data collection date.
2. Patient details: age, sex, name or national identity number
3. Exposure details: time and place, circumstances of exposure (occupational, intentional, and accidental), main activity at time of exposure (e.g., application in the field, household application, manufacturing, mixing, or loading), location of exposure (e.g., farm, field, home, urban, or rural). It should be noted that "accident" is the term used by the WHO Pesticide Project instead of the more commonly used "non-intentional."
4. Agent details: product identity (product name, active ingredient, chemical type, and physical form). WHO has a recommended categorization scheme for pesticides by hazard with guidelines for classification.[5]

5. Management details: medical treatment given, hospitalization (days in hospital), severity grading, poisoning severity score (none, minor, moderate, severe) and outcome (recovery, recovery with sequelae, death related, death unrelated). See later section for WHO Pesticide Poisoning Severity Score (PPSS) description.

DATA ANALYSIS

In order to allow a detailed analysis and interpretation of data collected on cases of acute poisoning by pesticides, the data collected should be comparable and compatible. To achieve this, data should be collected in a harmonized manner. What is needed includes standard definitions for case, case classification (suspected, or confirmed case), severity grading (none, minor, moderate, severe, or fatal) and clear instructions and training for form completion.

Guidance documents are available from the WHO Epidemiology of Pesticide Poisoning project,[6] DANIDA and the Pan American Health Organization PLAGSALUD project,[7] and the National Institute for Occupational Safety and Health (NIOSH) Pesticide Illness and Injury Surveillance Program.[8]

Additional data will be needed to analyze data which would include, but not be limited to, socio-demographic information, number of persons who have access to hospital or any other type of health care, major causes of death in the country, agricultural production information, amount of pesticides used, and type of pesticides.

FORMS DEVELOPED BY INTERNATIONAL ORGANIZATIONS

The Pesticide Exposure Record (PER) was developed through the activities of the WHO International Programme on Chemical Safety (IPCS) Epidemiology of Pesticide Poisoning project for collection of data in a harmonized format for participating countries. Also, the WHO IPCS has the INTOX Data Management System, which includes a communications record that is the format used for documenting enquiries to the poisons center.[9]

The secretariat for the Rotterdam Convention on the Prior Informed Consent Procedure for Certain Hazardous Chemicals and Pesticides in the International Trade has developed the Severely Hazardous Pesticide Formulation Report Form to facilitate the identification of candidate formulations for inclusion in the Rotterdam Convention.[10]

REFERENCES

1. Available from: www.pic.int [cited 2007 Feb 7].
2. Available from: www.paho.org/English/SHA/be_v23n3-plaguicidas.htm [cited 2007 Feb 7].
3. Available from: http://epi.minsal.cl/epi/html/normas/decreto88.htm [cited 2007 Feb 7].
4. Available from: www.fao.org/documents/show_cdr.asp?url_file = /docrep/008/af340e/af340e04.htm [cited 2007 Feb 7].
5. Available from: www.who.int/ipcs/publications/pesticides_hazard/en/ [cited 2007 Feb 7].
6. Available from: www.nihs.go.jp/GINC/meeting/7th/7report/doc3.pdf [cited 2007 Feb 7].
7. Available from: www.paho.org/English/SHA/be_v23n3-plaguicidas.htm [cited 2007 Feb 7].
8. Available from: www.cdc.gov/niosh/topics/pesticides/#reports [cited 2007 Feb 7].
9. Available from: www.intox.org [cited 2007 Feb 7].
10. Available from: www.pic.int [cited 2007 Feb 7].

doi:10.1300/J096v12n01_09

PESTICIDE POISONING SEVERITY SCORE (PPSS) IPCS Modification of the IPCS/EC/ EAPCCT PSS

A standardized scale for grading the severity of poisoning by pesticides allows qualitative evaluation of morbidity caused by poisoning, better identification of real risks and comparability of data.

INSTRUCTIONS

Poisoning Severity Score (PSS) is a classification scheme for cases of pesticide poisoning in adults and children. This scheme should be used for the classification of acute pesticide poisonings regardless of the type and number of agents involved. However, modified schemes may eventually be required for certain specific poisonings and this scheme may then serve as a model.

The PSS should take into account the overall clinical course and be applied according to the most severe symptomatology (including both subjective symptoms and objective signs). Therefore it is normally a retrospective process, requiring follow-up of cases. If the grading is undertaken at any other time (e.g. on admission) this must be clearly stated when the data are presented.

The use of the score is simple. The occurrence of a particular symptom is checked against the chart and the severity grading assigned to a case is determined by the most severe symptom(s) or sign(s) observed.

Severity grading should take into account only the observed clinical symptoms and signs and it should not estimate risks or hazards on the basis of parameters such as amounts ingested or serum/plasma concentrations.

The signs and symptoms given in the scheme for each grade serve as examples to assist in grading severity.

Treatment measures employed are not graded themselves, but the type of symptomatic and/or supportive treatment applied (e.g. assisted ventilation, inotropic support, hemodialysis for renal failure) may indirectly help in the evaluation of severity. However, preventive use of antidotes should not influence the grading, but should instead be mentioned when the data are presented.

Although the scheme is in principle intended for grading of acute stages of poisoning, if disabling sequelae and disfigurement occur, they would justify a high severity grade and should be commented on when the data are presented. If a patient's past medical history is considered to influence the severity of poisoning this should also be commented on.

Severe cases resulting in death are graded separately in the score to allow a more accurate presentation of data (although it is understood that death is not a grade of severity but an outcome).

SEVERITY GRADES

NONE (0): No symptoms or signs related to poisoning.
MINOR (1): Mild, transient and spontaneously resolving symptoms.
MODERATE (2): Pronounced or prolonged symptoms.
SEVERE (3): Severe or life-threatening symptoms.
FATAL (4): Death.

ORGAN	NONE	MINOR	MODERATE	SEVERE	FATAL
	0	1	2	3	4
	No symptoms or signs	Mild, transient and spontaneously resolving symptoms or signs	Pronounced or prolonged symptoms or signs	Severe or life-threatening symptoms or signs	Death
GI-tract		• Vomiting (<5/24hrs), diarrhoea (4 to 6/24hrs), pain, nausea • Mild salivation • Irritation, 1st degree burns, minimal ulcerations in the mouth • Endoscopy: Erythema, oedema	• Pronounced or prolonged vomiting, diarrhoea, pain, ileus • 1st degree burns of critical localization or 2nd and 3rd degree burns in restricted areas • Dysphagia • Endoscopy: Ulcerative transmucosal lesions	• Massive haemorrhage, perforation • More widespread 2nd and 3rd degree burns • Severe dysphagia • Endoscopy: Ulcerative transmural lesions, circumferencial lesions, perforation	
Respiratory system		• Irritation, coughing, breathlessness, mild dyspnea, mild bronchospasm • Bradypnoea <20/min • Chest X-ray: Abnormal with minor or no symptoms	• Prolonged coughing, bronchospasm, dyspnea, stridor, hypoxemia requiring extra oxygen • Chest X-ray: Abnormal with moderate symptoms	• Manifest respiratory insufficiency (due to, e.g., severe bronchospasm, airway obstruction, glottal oedema, pulmonary oedema, excessive bronchial secretions, ARDS, pneumonitis, pneumonia, pneumothorax) • Respiratory arrest • Chest X-ray: Abnormal with severe symptoms	
Nervous system		• Drowsiness, vertigo, tinnitus, ataxia • Restlessness • Mild extrapyramidal symptoms • Mild cholinergic/ anticholinergic symptoms • Miosis	• Unconsciousness with appropriate response to pain • Brief apnoea, bradypnoea • Confusion, agitation, hallucinations, delirium • Infrequent, generalized or local seizures • Pronounced extrapyramidal symptoms • Pronounced cholinergic/ anticholinergic symptoms • Miosis	• Deep coma with inappropriate response to pain or unresponsive to pain • Respiratory depression with insufficiency • Extreme agitation • Frequent, generalized seizures, status epilepticus, opisthotonus • Miosis (pin point pupils)	
Nervous system		• Paresthesia • Mild visual or auditory disturbances • Absence of fasciculations	• Localized paralysis not affecting vital functions • Visual and auditory disturbances • Isolated fasciculations • *Pronounced sweating*	• Generalized paralysis or paralysis affecting vital functions • Blindness, deafness • Generalized continuous fasciculations • *Extreme, generalized sweating*	

ORGAN	NONE	MINOR	MODERATE	SEVERE	FATAL
	0	1	2	3	4
	No symptoms or signs	Mild, transient and spontaneously resolving symptoms or signs	Pronounced or prolonged symptoms or signs	Severe or life-threatening symptoms or signs	Death
Cardio-Vascular system		• Heart Rate: 60 to 100 pm	• Sinus bradycardia (HR ~40-50 in adults, 60-80 in infants and children, 80-90 in neonates)	• Severe sinus bradycardia (HR ~<40 in adults, <60 in infants and children, <80 in neonates)	
			• Sinus tachycardia (HR ~140-180 in adults, 160-190 in infants and children, 160-200 in neonates)	• Severe sinus tachycardia (HR ~>180 in adults, >190 in infants and children, >200 in neonates)	
		• Isolated extrasystoles	• Frequent extrasystoles, atrial fibrillation/ flutter, AV-block I-II, prolonged QRS and QTc-time, repolarization abnormalities	• Life-threatening ventricular dysrythmias, AV block III, asystole	
			• Myocardial ischaemia	• Myocardial infarction	
		• Mild and transient hypo/hypertension	• More pronounced hypo/hypertension	• Shock, hypertensive crisis	
				• Cardiac arrest	
Metabolic balance		• Mild acid-base disturbances (HCO_3^- ~15-20 or 30-40 mmol/l, pH~7.25-7.32 or 7.50-7.59)	• More pronounced acid-base disturbances (HCO_3^- ~10-14 or >40 mmol/l, pH ~7.15-7.24 or 7.60-7.69)	• Severe acid-base disturbances (HCO_3^- ~<10 mmol/l, pH ~<7.15 or >7.7)	
		• Mild electrolyte and fluid disturbances (K^+ 3.0-3.4 or 5.2-5.9 mmol/l)	• More pronounced electrolyte and fluid disturbances (K^+ 2.5-2.9 or 6.0-6.9 mmol/l)	• Severe electrolyte and fluid disturbances (K^+ <2.5 or >7.0 mmol/l)	
		• Mild hyperglycaemia			
		• Mild hypoglycaemia (~50-70 mg/dl or 2.8-3.9 mmol/l in adults)	• More pronounced hypoglycaemia (~30-50 mg/dl or 1.7-2.8 mmol/l in adults)	• Severe hypoglycaemia (~<30 mg/dl or 1.7 mmol/l in adults)	
		• Hyperthermia of short duration	• Hyperthermia of longer duration	• Dangerous hypo- or hyperthermia	
Liver		• Minimal rise in serum enzymes (ASAT, ALAT ~2-5 × normal)	• Rise in serum enzymes (ASAT, ALAT ~5-50 × normal) but no diagnostic biochemical (e.g., ammonia, clotting factors) or clinical evidence of liver dysfunction	• Rise in serum enzymes (~>50 × normal) or biochemical (e.g., ammonia, clotting factors) or clinical evidence of liver failure	
Kidney		• Minimal proteinuria/haematuria	• Massive proteinuria/haematuria		
			• Renal dysfunction (e.g., oliguria, polyuria, serum creatinine of ~200-500 µmol/l)	• Renal failure (e.g., anuria, serum creatinine of >500 µmol/l)	
Blood		• Mild haemolysis	• Haemolysis	• Massive haemolysis	
		• Mild methaemoglobinaemia (metHb ~10-30%)	• More pronounced methaemoglobinaemia (metHb ~30-50%)	• Severe methaemoglobinaemia (metHb >50%)	
			• Coagulation disturbances without bleeding	• Coagulation disturbances with bleeding	
			• Anaemia, leucopenia, thrombocytopenia	• Severe anaemia, leucopenia, thrombocytopenia	
			• Inhibited Acetylcholinesterases	• Acetylcholinesterases severely inhibited	

ORGAN	NONE	MINOR	MODERATE	SEVERE	FATAL
	0	1	2	3	4
	No symptoms or signs	Mild, transient and spontaneously resolving symptoms or signs	Pronounced or prolonged symptoms or signs	Severe or life-threatening symptoms or signs	Death
Muscular system		• Mild pain, tenderness • CPK ~250-1,500 iu/l	• Pain, rigidity, cramping and fasciculations • Rhabdomyolysis, CPK ~1,500-10,000 iu/l	• Intense pain, extreme rigidity, extensive cramping and fasciculations • Rhabdomyolysis with complications, CPK ~>10,000 iu/l • Compartment syndrome	
Local effects on skin		• Irritation, erythema • 1st degree burns (reddening) or 2nd degree burns in <10% of body surface area	• Marked irritation, erythema, oedema • 2nd degree burns in 10-50% of body surface (children: 10-30%) or 3rd degree burns in <2% of body surface area	• Lesion • 2nd degree burns in >50% of body surface (children: >30%) or 3rd degree burns in >2% of body surface area	
Local effects on eye		• Irritation, redness, lacrimation, mild palpebral oedema	• Intense irritation, corneal abrasion • Minor (punctate) corneal ulcers	• Corneal ulcers (other than punctate), perforation • Permanent damage	

For further information, copies and comments, turn to:

Dr Jenny Pronczuk de Garbino or *Dr Hans Persson*
IPCS/PHE/WHO   *Swedish Poisons Information Centre*
CH-1211 Geneva 27, Switzerland   *S-171 76 Stockholm, Sweden*
Fax: +41 22 791 4848   *Fax: +46 8 327584*
Tel: +41 22 791 3602/3595   *Tel: +46 8 6100511*
E-mail: pronczukj@who.ch   *E-mail: hans.persson@apoteket.se*

Index

Note. Page numbers followed by the letter "t" designate tables; numbers followed by the letter "f" designate figures.

Abortions. *See* Spontaneous abortions, exposure to pesticides and
Acute contact dermatitis, 3
Acute irritant contact dermatitis, 4
Acute pesticide intoxication, 1
Agency for Toxic Substances and Disease Registry (ATSDR), 80
Agent Orange (2,4,5-T), 69
Agricultural chemicals, TRUE test and, 5
Agricultural Health Study (AHS), 20,30,81
Agricultural workers, 8
Aldicarb, 19
Aldrin, 18,69
Allergic contact dermatitis, 4
  carbamates and, 6-7
  causes of, 4
  diagnostic testing for, 4-5
Amitraz, 20
Animal surrogates, risk assessment of pesticides, 39-40
Antimicrobials, 7
Arsenic, 22
Arsenical insecticides, 40
Assays, skin irritation, 6
Association of Occupational Environmental Clinics, 81
Atrazine, spontaneous abortions and exposure to, 31
ATSDR. *See* Agency for Toxic Substances and Disease Registry (ATSDR)
Azinphos methyl, 19

Benlate (benomyl), 7
Benomyl (Benlate), 7
Biological insecticides, 7
Biological monitoring, 41
Biomarkers, cancers and, 47-48
Birth defects, exposure to pesticides and
  among men, 29-30
  among women, 31
Breast milk contamination, exposure to pesticides and, 31-32

California Department of Pesticide Regulation (DPR), 78,79
California Environmental Protection Agency, 79

California Pesticide Illness Surveillance Program (PISP), 58-62
California Pesticide Use Reporting database, 41-42
Cancers
  adult, 45
    colorectal, 43
    leukemia, 44
    lung, 42-43
    multiple myeloma, 44
    non-Hodgkin's lymphoma, 44
    pancreatic, 43-44
    prostate, 42
    soft-tissue sarcoma, 45
  biomarkers of early effect, 47-48
  childhood, 45-47
  linking pesticides with, 47
  susceptibility factors from pesticides for, 48
Capsicum oleoresin-capsaicin, 7
Captan, 7
Captofol, 7
Carbamate insecticides, 6-7
Carbamates, 7
Carbaryl, 19
  as risk factor for non-Hodgkin's lymphoma, 44
Carcinogencity, human
  exposure assessment for, 40-42
  pesticides and, 39-40
Carpet dust, as medium for environmental monitoring, 41
Carson, Rachel, 69
CCA. *See* Copper-chromated-arsenate (CCA)
CDC. *See* Centers for Disease Control and Prevention (CDC)
Center for Environmental and Regulatory Information Systems (CERIS), Purdue University, 78
Centers for Disease Control and Prevention (CDC), 2,80
Central nervous system. *See also* Peripheral nervous system
  chloinesterase inhibitors and, 19
  dysfunction symptoms of, 18
  fumigants and, 19-20
  novel agents and, 20
  organochlorine pesticides and, 18-19
  Parkinson's Disease and, 20

Available online at http://ja.haworthpress.com
© 2007 by The Haworth Press, Inc. All rights reserved.

Printed and bound by CPI Group (UK) Ltd, Croydon, CR0 4YY

22/10/2024

01777634-0015